A New Era for Collaborative Forest Management

This book assesses the Collaborative Forest Landscape Restoration Program (CFLRP) and identifies lessons learned for governance and policy through this new and innovative approach to collaborative forest management.

Unlike anything else in US public land management, the CFLRP is a nationwide program that requires collaboration throughout the life of national forest restoration projects, joining agency partners and local stakeholder groups in a kind of decade-long restoration marriage. This book provides a comprehensive assessment of the governance dynamics of the program, examining: questions about collaborative governance processes and the dynamics of trust, accountability and capacity; how scientific information is used in making decisions and integrated into adaptive management processes; and the topic of collaboration through implementation, an underdeveloped area of collaborative governance literature. Bringing together chapters from a community of social science and policy researchers who have conducted studies across multiple CFLRP projects, this volume generates insights, not just about the program, but also about dynamics that are central to collaborative and landscape approaches to land management and relevant for broader practice.

This volume is a timely and important contribution to environmental governance scholarship. It will be of interest to researchers and students of natural resource management, environmental governance, and forestry, as well as practitioners and policy makers involved in forest and ecosystem restoration efforts, and collaborative natural resource management more broadly.

William H. Butler is Associate Professor of Environmental Planning in the Department of Urban and Regional Planning at Florida State University, USA. His research has focused on collaborative environmental planning and management, particularly in forest management, ecological restoration, and climate change adaptation.

Courtney A. Schultz is Associate Professor of Forest and Natural Resource Policy and Director of the Public Lands Policy Group in the Department of Forest and Rangeland Stewardship at Colorado State University, USA. Her research focuses on the intersection of science, policy, and law in natural resource management, particularly how policy mediates the adoption and application of science-based land management principles.

The Earthscan Forest Library

This series brings together a wide collection of volumes addressing diverse aspects of forests and forestry and draws on a range of disciplinary perspectives. Titles cover the full range of forest science and include the biology, ecology, biodiversity, restoration, management (including silviculture and timber production), geography and environment (including climate change), socio-economics, anthropology, policy, law, and governance. The series aims to demonstrate the important role of forests in nature, peoples' livelihoods and in contributing to broader sustainable development goals. It is aimed at undergraduate and postgraduate students, researchers, professionals, policy-makers and concerned members of civil society.

Series Editorial Advisers:

Recent Titles:

Forests for People
Community Rights and Forest Tenure Reform
Edited by Anne M. Larson, Deborah Barry, Ganga Ram Dahal and Carol J. Pierce Colfer

Logjam
Deforestation and the Crisis of Global Governance
David Humphreys, with a foreword by Jeffrey Sayer

The Decentralization of Forest Governance
Politics, Economics and the Fight for Control of Forests in Indonesian Borneo
Edited by Moira Moeliono, Eva Wollenberg and Godwin Limberg

Additional information on these and further titles can be found at www.routledge.com/books/series/ECTEFL

A New Era for Collaborative Forest Management

Policy and Practice Insights from the Collaborative Forest Landscape Restoration Program

Edited by William H. Butler and Courtney A. Schultz

Routledge
Taylor & Francis Group
LONDON AND NEW YORK

earthscan
from Routledge

First published 2019
by Routledge
2 Park Square, Milton Park, Abingdon, Oxon OX14 4RN

and by Routledge
52 Vanderbilt Avenue, New York, NY 10017

First issued in paperback 2020

Routledge is an imprint of the Taylor & Francis Group, an informa business

British Library Cataloguing-in-Publication Data
A catalogue record for this book is available from the British Library

Library of Congress Cataloging-in-Publication Data
A catalog record has been requested for this book

ISBN 13: 978-0-367-66274-5 (pbk)
ISBN 13: 978-1-138-49127-4 (hbk)

Typeset in Bembo
by Wearset Ltd, Boldon, Tyne and Wear

C.S. for my family and husband, with gratitude for their unfaltering support.

W.B. for my wife, Rachel, and boys, Alexander and Evan, with deep appreciation for their affirmations to see this through.

Contents

Contributors

Gregory H. Aplet is the Senior Science Director at the Wilderness Society in Denver, where he supervises a team of ecologists and geographic information professionals located in six regional offices throughout the country.

Hannah A. Bergemann is a graduate student at Colorado State University pursuing an MS in Forest Science under Dr Courtney Schultz. She is employed with the US Forest Service through the Pathways Internship Program, where she works for the Washington Office's Forest Management unit out of Fort Collins, CO.

Susan Jane M. Brown is a staff attorney with the Western Environmental Law Center (WELC). Her primary focus of litigation is federal public lands forest management, but her practice includes cases involving the Endangered Species Act, the National Environmental Policy Act, the National Forest Management Act, and other land management statutes. She is a former co-chair of the National Advisory Committee for Implementation of the National Forest System Land Management Planning Rule, and is also heavily engaged in collaborative forest restoration in eastern Oregon.

William H. Butler is an Associate Professor of Environmental Planning in the Department of Urban and Regional Planning at Florida State University. Will's research focuses on collaborative environmental planning and management particularly in: forest management and ecological restoration; sea level rise and climate adaptation; and food systems planning. He has led a research project examining the transition from collaborative planning to implementation since the inception of CFLRP.

Antony (Tony) S. Cheng is the Director of the Colorado Forest Restoration Institute and Professor in the Department of Forest & Rangeland Stewardship at Colorado State University. Tony's primary research and applied work is in forest governance, policy, and administration, with a focus on collaborative approaches to promote resilient social-ecological systems linked to forest landscapes.

Rachel A. Christensen is a Research Associate with the Institute for Policy and Governance and adjunct professor of non-profit studies at Virginia Tech. Her research explores how non-profits contracting and collaborating with government agencies navigate tensions across multiple dimensions of accountability.

Melanie M. Colavito is the Human Dimensions Specialist for the Ecological Restoration Institute at Northern Arizona University in Flagstaff. Her work focuses on the development, communication, and application of use-inspired scientific information for collaborative forest management. She earned a PhD in geography with a minor in remote sensing and spatial analysis from the School of Geography and Development at the University of Arizona in Tucson.

Kimberly Coleman is an assistant professor in the Center for Earth and Environmental Science at SUNY Plattsburgh. She received a BS and a MS from the University of Vermont in environmental studies and natural resources, respectively, as well as a PhD in forest resources and environmental conservation from Virginia Tech. Her research focuses on questions related to collaborative forest planning and management, sustainability education, equity, and cultural ecosystem services.

Emily Jane Davis is an assistant professor and extension specialist in the Department of Forest Ecosystems and Society at Oregon State University. She is a qualitative social scientist with a focus on social and economic aspects of natural resource management, policy, and practice. Her applied research and technical assistance involves meeting and process design, monitoring protocols, strategic planning, facilitation, grant reviews, partnership development, program and project evaluation, and workshop planning. She holds a PhD and MA from the University of British Columbia and a BA from McGill University.

Bryce Esch works at the Ecological Restoration Institute at Northern Arizona University where she provides science support for forest restoration efforts throughout the western US. Bryce has been studying collaborative forest restoration since 2012 and working on ecological restoration and monitoring with the Four Forest Restoration Initiative since 2015. She received her MS in Conservation Ecology from the School of Natural Resources and Environment at the University of Michigan.

Alexander Heeren earned his PhD from The Ohio State University's School of Environment and Natural Resources. He is currently a post doc at the University of California Merced's Sierra Nevada Research Institute. His research focuses on how individuals perceive natural resources, sustainability, and environmental change.

Richard D. Margerum, is Professor and Head of the School of Planning, Public Policy, and Management at the University of Oregon. He researches collaborative approaches to governance with a particular emphasis on metropolitan regions and natural resources management. He is the author of over twenty-five referred publications, including his book: *Beyond Consensus: Improving Collaborative Planning and Management* (MIT Press, 2011).

Sarah McCaffrey is a research social scientist for the Rocky Mountain Research Station of the USDA Forest Service. She conducts and coordinates research to better understand the social dynamics of fire management. She has initiated dozens of studies in a variety of ecological and geographical settings. She has published numerous peer-review articles as well as research reports oriented to practitioner audiences in this field.

Laura McCarthy is Associate State Director for the Nature Conservancy in New Mexico. She has been with the Conservancy since 2005, including seven years as the Conservancy's National Policy Advisor for fire management and forest restoration. Laura enjoys her role building bridges that connect people to nature, including land and water managers, and urban, rural, and traditional cultural communities. She has earned several awards for her work including New Mexico Environmental Leader of the Year in 2015 and the Yale School of Forestry and Environmental Studies Distinguished Alumna Award in 2017.

Kathleen B. McIntyre is a PhD candidate at Colorado State University in the Department of Forest and Rangeland Stewardship, focusing on public lands policy and collaborative governance. Kathleen has a BS from the University of Michigan in Environmental Science and an MS from the University of Michigan in Environmental Policy and Planning.

Ashley Monroe graduated with her MSP and PhD in Environmental Planning from Florida State University. Her research focused on collaborative forest restoration processes and how to structure collaborative efforts for effective implementation using CFLRP cases for her work. She works for the Environmental Protection Agency evaluating stream and wetland mitigation projects and serving on the steering committees of a number of the Landscape Conservation Cooperatives.

Clare M. Ryan is a Professor of Natural Resource and Environmental Policy at the University of Washington's School of Environmental and Forest Sciences in the College of the Environment. Her research and teaching focus on the processes by which policy is made and implemented, ways to foster collaboration in resource management, how to address conflicts that can arise when multiple stakeholders participate in decision-making, and how scientific information can be integrated into policy and management decisions.

Courtney A. Schultz is Associate Professor of Forest and Natural Resource Policy in the Department of Forest and Rangeland Stewardship at Colorado

State University. She studies US national forest policy to inform practice and broader scholarship on policy design to support improved environmental governance. She has written extensively on wildlife conservation planning, adaptive management, forest restoration, and climate change adaptation strategies on US public lands, and has been conducting research on the CFLRP since its inception. Courtney directs the Public Lands Policy Group (PLPG) at Colorado State University, which produces original research to inform the practice of natural resource management and advance understanding of policy developments that affect US public lands.

Marc J. Stern is a Professor in the Department of Forest Resources and Environmental Conservation at Virginia Tech. His research, teaching, and outreach focus on the predictors and consequences of human behaviour across multiple contexts, including environmental decision-making, collaborative natural resource management, and protected areas management. He is also the author of *Social Science Theory for Environmental Sustainability: A Practical Guide*.

Eric Toman is an Associate Professor and Chair of the Graduate Studies Committee in the School of Environment and Natural Resources at The Ohio State University. His research explores human behaviours in the context of environmental and social change and considers how social, psychological, and contextual variables interact to influence the resilience of social and ecological systems.

Lauren S. Urgenson is an interdisciplinary Research Scientist at the University of Washington's School of Environmental and Forest Sciences in the College of the Environment. Her work focuses on ecological and social dimensions of forest ecosystem science, policy, and management. Lauren was awarded a PhD in Ecological Restoration, a Masters in Forest Ecology from the University of Washington, and a BA in Environmental Science and Policy from the University of Maryland.

Emily H. Walpole is a PhD candidate in the School of Environment and Natural Resources at The Ohio State University. She earned her MS in 2015 and is currently investigating climate adaptation perceptions and decision-making as part of her dissertation research. She also serves as the manager of the Environmental and Social Sustainability (ESS) Lab at Ohio State.

Amy E.M. Waltz is Director of Science Delivery and a research ecologist with the Ecological Restoration Institute (ERI) at Northern Arizona University. Amy works to translate and deliver peer-reviewed science relevant to landscape-scale ecological restoration to land managers and multi-stakeholder groups.

Preface and acknowledgements

When the Collaborative Forest Landscape Restoration Program (CFLRP) was created in 2009, many researchers viewed the program as a major innovation in forest management policy. Shortly after the first round of funding was awarded to ten landscape restoration projects, we both were fortunate enough to begin research on the CFLRP projects and collaborative groups, and many of the contributors in this book also started work along parallel lines of inquiry.

After about a year of working independently from each other, we began to recognize the complementarity (and potential overlap) of our collective work and convened a group of researchers, including many of the contributors of chapters in this volume, in periodic conference calls to share research ideas and insights. Our goals were to minimize repetition and research fatigue among CFLRP participants, and also to share findings and discuss emerging insights. We collectively developed a community of scholars who were interested in landscape-scale restoration and collaborative forest management. Along the way, partners from the Nature Conservancy and National Forest Foundation joined in on our conversations, as did CFLRP program leaders and researchers from the US Forest Service, to help us think more critically about our research and to connect us directly to the CFLRP practitioner community.

During the later years of the CFLRP, we recognized the growing synergies in the research that scholars were conducting around the country and the need to consolidate the work on such an important change to forest management policy and governance. The research was simultaneously theoretically rich and deeply tied to policy and practice. We sought to bring this work together to capture the cross-cutting insights in one place – this book – and also to enhance our collective work through a cross-pollination of ideas and critical reflection.

To accomplish this, in the fall of 2017, we brought the scholars included in this volume together in a symposium, along with senior practitioners from several organizations (e.g. Rural Voices for Conservation Coalition, US Forest Service, and the National Forest Foundation) participating in CFLRP projects. We also invited several other scholars, some of whom provide commentary in this volume, with less direct ties to CFLRP, but who could critically analyse our work in collaborative forest management. The symposium was held at Florida State University, where we spent two days together, presenting initial versions of chapters and holding workshop discussions to identify and refine policy and practice implications reflected in the chapters, commentaries, and the concluding chapter of this book.

Beyond this gathering, which included opportunities for feedback from scholars and practitioners involved in the field, we also undertook an internal peer-review process for each of the chapters. Chapters were reviewed first by one of the editors of the book and then by another author included in the volume to ensure cross-pollination of ideas and to enhance the quality of the final chapters. The authors of the commentaries on each section provided a peer-review for at least one of the chapters and read and reviewed all of the chapters in their respective section to produce their commentaries. The editors then conducted a second and third round of reviews and requests for edits on all of the chapters. This process of review not only accomplished the desired cross-pollination of ideas, but it also helped identify potential inconsistencies, gaps, or other issues in the chapters.

It is a challenge to write about specific cases of collaborative forest management. These cases are highly dynamic and change over time. At best, researchers are able to take a series of snapshots along the path of collaborative restoration work. As a result, we sometimes miss some of the nuances and details of the collaborative group, participant interactions, and the activities and products in each case. Our assessment and evaluation of the cases is based on the most complete data we were able to gather as part of larger research projects. Each of the chapters is based on extensive qualitative field work and, occasionally, survey work to gather facts, experiences, actions, consequences, and perceptions. However, there is little doubt that, either due to changes over time or differing interpretations of activities, attitudes and perceptions among the varied participants in CFLRP projects, we will have made some claims that others might debate. We are certain there are people involved in many of these cases who are more familiar with the details and nuances, who will be able to think of contradictory evidence or differing interpretations to what we claim in our descriptions and analysis of CFLRP landscape projects and collaborative groups. We also recognize that we missed the perspective of people not participating in these projects, but who may have been observing them nonetheless. Any discrepancies in judgment, interpretation, or fact is the responsibility of the respective authors of each chapter. Our interpretations are inherently limited by the scope of the work we were able to undertake, and we invite continued conversations about these ideas, experiences, and interpretations to further refine them. Nonetheless, we remain confident that even if the particulars can be debated or contradicted, the broader insights herein remain relevant and important to moving theory, policy, and practice in collaborative forest management.

An edited volume does not come together without the commitment and support of many people. We would like to acknowledge as many as we can remember here, knowing that we may not thank everyone who deserves it. First, on behalf of all of the authors included in this volume, we would like to thank the CFLRP collaborative participants and Forest Service agency staff who contributed to our research. Taken together, the chapters herein drew on many hundreds of hours of interviews, surveys, and observations – time graciously given by people working to achieve restoration goals in their own part of the forest. The collective recognition that the CFLRP was a novel policy experiment prompted many people with too many demands on their time to pause and engage with all of us as we sought to capture what they were learning first-hand. We cannot express sufficiently our gratitude to each of them for taking time to share their experiences with us.

A handful of individuals and organizations have contributed funding to research on the CFLRP and supported this work through their willingness to connect us with project participants and share our work. The US Forest Service, and the Forest Management, Rangeland Management, and Vegetation Ecology staff, in particular, consistently provided financial and organizational support for the research that led to this book. Beyond this, throughout the CFLRP's years of implementation they consistently sought feedback from the research community and from practitioners in the field. We want to thank, in particular, CFLRP Program Coordinator Lindsay Buchanan and Integrated Restoration Coordinator Jessica Robertson for their support of the research and their dedication to always seeking feedback, ideas, and strategies for supporting and improving collaborative forest restoration efforts. From the beginning, Forest Service research scientist Sarah McCaffrey recognized the value of investing in research on this program and supported several projects from the US Forest Service Northern Research Station. The McIntire-Stennis Cooperative Forestry Research Program provided substantial funding, in combination with institutional support at Colorado State University, for many years of work on the CFLRP. Other funders of research in this volume included the Marshall Foundation, the Redd Center for Western Studies, the National Institute for Food and Agriculture Hatch Funds, the Nature Conservancy's US Fire Learning Network, the National Forest Foundation, the US Forest Service Pacific Northwest Research Station, and the Institute of the Environment. The work on the CFLRP in this volume and elsewhere would not have been possible without research funding from these entities and others. This financial support always came with an understanding of the need for research and editorial independence, even when our work unearthed unflattering findings and failures. We have been consistently impressed with the broad interest among the community of practice in critical reflection on how to improve policy and practice going forward.

Bringing the authors together with high-level practitioners from the field for the symposium in the fall 2017 helped bring this book to fruition. We could not have held the symposium without the financial and staff support of the College of Social Sciences and Public Policy at Florida State University, the Department of Urban and Regional Planning at Florida State University, the Department of Forest and Rangeland Stewardship at Colorado State University, the US Forest Service, the Ecological Restoration Institute of Northern Arizona University, the Colorado Forest Restoration Institute, the National Forest Foundation, and the Rural Voices for Conservation Coalition. The symposium accomplished much more than a meeting of the minds – it led to fruitful and creative idea generation that has influenced the content of this book as well as informed work toward the next iteration of landscape level forest restoration policy.

We would also like to thank the authors included in this volume. We were honoured to bring together this illustrious group and were pleased with the contributions the authors made, not only in developing their own chapters in a timely fashion, but also contributing to the peer review process to ensure cross-pollination of ideas and quality of contributions. It is our privilege to be part of this community of scholars.

Finally, we wish to thank the staff at Routledge Earthscan for their interest in applied scholarship and for their input, editing, and guidance as we navigated the process of bringing this book to press.

Foreword

Laura McCarthy

The week between Christmas and New Year is a time I enjoy for a few days of professional reflection. The flow of email slows to a trickle and the office is quiet. At the end of 2006, I was focused on the US Congress and preparing materials for the new Chair of the Energy and Natural Resources Committee, New Mexico Senator, Jeff Bingaman. Two projects were top of my mind: Southwest Jemez in New Mexico and Tapash in Washington, both poised to restore forests at an unprecedented scale, and both stymied by the same factors. I took a few days and wrote a brief concept paper where I proposed the creation of a federal priority landscapes program to restore fire-adapted forests.

My central argument was that investment in ecological treatments and economic development to restore forests and communities would accomplish more if the funding were targeted at places that were well-positioned to achieve immediate and tangible results. Rather than restore a little bit everywhere, the objective was to direct a percentage of existing federal funding to large landscapes where a targeted investment would make a significant difference. When everything is a priority, nothing is a priority, and scientific knowledge of fire-adapted ecosystems makes clear that fire behaviour will not change unless forests are restored on a significant scale.

I was new to the legislative process. My credentials came from on-the-ground experience in community-based forestry in New Mexico and from my engagement with the Rural Voices for Conservation Coalition (RVCC), a group of community-based practitioners advocating for policy changes to support the rural West. Fortunately, my colleagues at The Nature Conservancy, Michael Powelson and Tom Cassidy, brought to our team knowledge of how legislation is developed. They transmitted my concept paper to staff at the Energy and Natural Resources Committee, and so began the drafting process.

Senator Bingaman was the ideal sponsor for potential legislation because of his prior work in 2000 to create the Community Forest Restoration Act (Title VI, Public Law 106–393) that funded collaborative, community-based, ecological restoration projects on public lands in New Mexico. As the chair of the Senate Energy and Natural Resources Committee, the Senator embraced the idea of using the lessons learned from New Mexico to create a nationwide program. The ranking member on the Committee, Senator Pete Domenici, also from New Mexico, was interested as well. Senator Domenici was seeking to increase active management of the national forests and break through gridlock over timber harvests. The third sponsor, California Senator Dianne Feinstein, came to the table because of devastating wildfires in California and the decline of a once-robust federal timber program, which was disrupting the economies of many rural communities. As chair of the Interior Appropriations Committee, her engagement helped assure that if legislation were authorized, it could be funded.

The bill was drafted in 2007 and introduced in 2008, with input from many groups and individuals working on Capitol Hill. Community foresters, forest industry representatives, and wilderness advocates, among others, provided input and feedback on the draft language. Their comments generated significant changes to the drafts; for example, the multi-party monitoring provisions were included because of advocacy by RVCC (and have turned out to be a key aspect of the program). The final version of the bill was included in the Omnibus Public Land Management Act of 2009, and when it was passed in the first 100 days of the Obama Administration by a Democratic Congress as the Forest Landscape Restoration Act of 2009 (P.L. 11–111; 16 U.S.C. §7303), the Collaborative Forest Landscape Restoration Program (CFLRP) was created. We celebrated the legislation, and I thought that the hard work was done.

Securing an appropriation for the CFLRP took even greater effort. Having fostered an inclusive process to review and revise the legislation, we marshalled the groups that had worked on the bill to form a coalition to support funding. We generated letters and created fact sheets, using the Southwest Jemez and Tapash projects as the examples. A few months after the bill was passed, Congress appropriated $10 million in the Fiscal Year (FY) 2010 Interior Appropriation Bill – enough to launch the program. Ramping up funding for CFLRP to the fully authorized level of $40 million took an additional two years, a much larger coalition, more fact sheets, and even greater effort. By FY 2012, the full authorized funding amount of $40 million was secured.

Meanwhile, the Forest Service was gearing up to implement the new program. Another surprise ensued – Forest Service staff were interpreting the legislation differently than we had imagined. Our outreach to key officials to explain the legislation's intent was rebuffed. Suffice to say, the Forest Service wanted to set up the new program on their own. We had no choice but to sit on the sidelines and watch how it played out. The law required new approaches from the Forest Service: the FLRA was the first law enacted to establish a national program requiring a collaborative process as a prerequisite for submitting a proposal for funding, and to establish a competitive selection process based on recommendations from a federal advisory committee of diverse stakeholders. These features were not entirely new, they had been tested in the New Mexico Collaborative Forest Restoration Program (CFRP). But, the experienced staff from New Mexico were not invited to participate in setting up the program, so the lessons learned were not carried forward.

With hindsight, the two major differences between our expectations for the law's implementation and the reality of how the program was carried out don't as seem surprising as they did at the time. First, we had expected that with a $40 million authorization the agency would select ten projects, each with an annual budget of $4 million. Instead, the agency selected ten projects when the appropriation was at $10 million, and then ramped it up to twenty-three projects when the appropriation hit $40 million. The logic was sensible: the Forest Service anticipated that projects would need either more or less funding in different years, and they didn't want unspent program balances. If the total amount needed among the projects for any given year exceeded $40 million, the agency could pull from other budget line items with similar purposes to fill any gaps. Their logic had another advantage: twenty-three projects meant projects in more states and congressional districts, and that helped to build broad support in Congress for the program.

The second difference was more consequential in nature and a result of needs that only became apparent after a few years of program implementation. As I mentioned earlier, the law required that a federal advisory committee be convened to review project applications and make recommendations for project funding. My expectation – which today I might attribute to magical thinking – was that the committee would be run like New Mexico's CFRP advisory panel. That is, I expected that a shared learning process would emerge between the Forest Service, project proponents, and committee members. It didn't happen. The review process was tightly controlled by the Forest Service and an environment for idea exchange was not fostered. Soon after the selection process was over, the advisory committee was disbanded, whereas it could have played a valuable, ongoing role for program adjustment, oversight, and transparency.

Luckily, shared learning has emerged in other ways, often at the project scale instead of nationally, through regional meetings of CFLRP group participants and through research sponsored by the USDA Forest Service. Right at the start, Sarah McCaffrey at the Rocky Mountain Research Station had the foresight to sponsor a multi-year study of the projects led by the co-editor of this volume, William Butler. Closer to the end of the first ten years of the program, the Forest Service's National Forest System funded a comprehensive evaluation of the program, led by Courtney Schultz, to add a third-party perspective on the program to their own, ongoing internal reviews. The editors and authors of this volume have captured many of the important lessons from this experimental approach to collaborative management of large forested landscapes. Three findings in this book resonate especially strongly for me: the strength of the collaborative experience; the importance of ten years of sustained funding; and the unexpected benefits of working at a larger scale.

Building deeper collaborative experience

When the CFLRP was launched, collaboration between stakeholders and national forest managers was the exception, not the rule. Examples of collaborative success existed in a few places, especially in areas where RVCC was active in the Pacific Northwest and in the southwestern states. Collaboration in these places was focused on planning, and, consequently, it was fragile and dependent on the motivation and personalities of line officers. Projects often disintegrated when key personnel changed positions, funding levels changed, or collaborative work transitioned from planning to implementation. Outcomes from these collaboratives were sporadic. As the findings in this volume demonstrate, with ten years of CFLRP implementation and mandated collaboration most people reported improved levels of trust, transparency, accountability, and capacity. This has the potential to be a game changer for the agency, building a more widespread culture of acceptance of and capacity to engage more effectively in collaborative forest management.

Long-term funding commitments

In the years leading up to the CFLRP, community forestry practitioners noticed that inconsistent funding levels were hindering the implementation of forest restoration. Funding was frequently moved to new projects in different areas, and project duration was too short to build local capacity in areas of collaboration, monitoring, workforce development, and product utilization. The theory behind the CFLRP was that consistent funding for ten years would provide sufficient time to develop partner and workforce capacity, facilitate learning and innovation, and build stronger relationships.

Consistent funding was also expected to allow time for projects to demonstrate a reliable supply of wood and biomass and, therefore, to stimulate new investment in wood processing. While capacity increased and relationships improved, the economic benefits of wood utilization have not materialized. That remains a challenge to be resolved. However, stronger relationships and higher capacity have benefits that can be transferrable, and new problems that arise have the potential to be less daunting when collaborative groups and the agency are working together more effectively. CFLRP was also intended to provide managers with new flexibility to use funds for a suite of restoration activities instead of being tied to specific activities. For example, if a field crew was travelling to a location to remove trees, they could also replace a road culvert on the way in and install new stream structures at the site, providing efficiency that was hindered by traditional line-item appropriations.

As will be demonstrated in several chapters of this book, it is clear that the funding commitment and flexibility brought people to the table with a more robust commitment, and created opportunities to increase efficiencies in restoration work. However, the funding is still time-bound. Understandably, a question that remains is whether restoration projects can be maintained once CFLRP funding ends. Losing funds after ten years of consistency is difficult, and from that perspective the concern is logical. But, if the project has not been able to lower the per-acre costs, nor offset costs with forest products or other revenue streams, and if the Forest doesn't consider the project a priority to continue with appropriated dollars then, perhaps, it is indeed time for activity to taper off. On the other hand, if a project resulted in increased capacity and improved efficiency, and the collaborators were successful at leveraging other funding, then it would stand to reason that the loss of CFLRP funds would be cushioned and the work would continue. On this point, only time will tell.

Increasing the pace and scale of restoration

As a graduate student in the late 1980s I was fascinated with the intersection of ecosystem ecology and silviculture. I observed that forest management was operating at a scale of hundreds of acres (often referred to as a "stand"), while natural processes like hurricanes and wildfires disturbed much larger areas. This disconnect bothered me, but with the CFLRP we had a chance to put into policy a requirement that the scale of management be matched to the scale of the disturbance. We had many discussions during the drafting process about the minimum landscape size. We settled on 50,000 acres, which could work in the west due to the high cost, extent, and severity of wildfires, and in the east where fragmented landscapes might be harder to piece together at much larger extents. While all landscape projects covered significantly larger areas than our minimum, the intent of the Act was met as landscape planning scales and the pace of restoration seem to have increased under CFLRP.

An unexpected finding reported in several of the chapters of this book was that the larger-landscape focus, multi-year funding commitment, and collaborative involvement gave projects the social license to experiment with larger NEPA planning areas that incorporated more types of restoration work. I also saw non-CFLRP projects voluntarily taking on the collaborative approach at a landscape-scale, and leveraged funding model of the CFLRP. I experienced it first-hand in New Mexico where at least five new large landscape collaborative projects were using the CFLRP model without the funding, but adopting elements of the program. In fact, this spill-over effect may turn out to be the greatest contribution of CFLRP to environmental governance because it signifies the evolution of the agency toward more formal levels of co-management.

Reauthorization of the CFLRP is pending in the Farm Bill as I write. We don't know yet if the CFLRP will continue for another ten years, if the twenty-three projects currently funded might be eligible for additional funding, or how many new projects would begin. In a workshop I attended about conservation finance in the winter of 2018, the question arose of whether to continue the Southwest Jemez project if CFLRP is reauthorized. Like many of the other CFLRP projects, the Southwest Jemez will begin its tenth year with 22,000 acres newly ready for treatment, with a recently constructed pellet mill nearby. But the price-per-acre of thinning is still too low for the small-diameter trees to pay their way out of the woods. At the workshop, partners were considering other revenue streams that might supplement the limited income from tree thinning. If CFLRP is reauthorized, and even if Southwest Jemez could reapply, would another ten years make sense? The partners were adamant that a strong value proposition had been created, and they wanted to fill their funding gap with new funding streams from cities and counties that benefitted from the project, through fire protection, clean water, and outdoor recreation. The commitment of partners to seeing the work move forward, with or without federal funding, and their desire to see the work continue without CFLRP dollars, is a good sign that CFLRP has set a strong foundation for continued work on at least some of the projects.

The CFLRP has shown us a path forward in this era of declining resources and escalating environmental challenges. It was a wholly unique policy experiment that helped us understand changing practice as we cope with increasing complexity in public lands management. The work in this volume captures the lessons learned, and points to the policy and practice implications and theoretical contributions for collaborative environmental governance. Regardless of whether the program continues, understanding the lessons learned and capturing the experience of these projects is an essential component of the ongoing, collective project of fostering collaborative approaches to forest management that can support work across landscapes to manage fire, water, commodity production, and recreation opportunities within and surrounding US public lands.

Introduction – the changing landscape of collaborative forest restoration

Courtney A. Schultz and William H. Butler

For forty-one days in early 2016, the nation was captivated by the armed occupation of the Malheur National Wildlife Refuge in Harney County, Oregon by a group of anti-government militiamen. Led by Ammon Bundy, an automobile fleet manager from Idaho with family roots in public lands ranching, the Malheur Refuge occupiers' primary objective was to highlight what they saw as 'government overreach' and excessive regulation of traditional uses of public lands such as logging, grazing, and mining. The occupiers demanded that the federal government turn over the Refuge and other public lands to Harney County as the rightful and true 'original' owner of the lands – a complicated argument given that those lands had been the traditional homeland of the Burns-Paiute Tribe .

Although many in Harney County and the surrounding communities sympathized with the occupiers' complaints about federal land management, the occupiers did not find a hospitable environment in which to recruit for their cause. Instead, residents, law enforcement, and elected officials repeatedly asked the occupiers to return home and leave the community in peace to work out their differences – including about how adjacent public lands should be managed – among themselves.

The people of this large but sparsely populated county (cows outnumber people fourteen to one) had a long history of working together to resolve conflicts related to natural resources management. The Steens Mountain Advisory Council, formed in 2000 as a result of federal legislation protecting Steens Mountain as a Cooperative Management and Protection Area, worked together to manage and restore almost 430,000 acres of federal land including a designated Wilderness Area. The High Desert Partnership – consisting of ranchers, federal land managers, conservationists, and others – formed in 2006 in response to local discontent with federal land management, had helped to write the management plan for the Wildlife Refuge under occupation. And the Harney County Restoration Collaborative, one half of the Southern Blues Restoration Coalition Collaborative Forest Landscape Restoration project, had been working together to restore thousands of acres of ponderosa pine forest on the south end of the Malheur National Forest since 2008. While imperfect, collaborative conservation, not confrontation, was the way in which residents preferred to resolve disputes involving federal land management.

This spirit of collaborative conservation runs throughout the Collaborative Forest Landscape Restoration Program (CFLRP) – the subject of this volume. We compiled this collection of work on the CFLRP because it is emblematic of a new era in US environmental policy. Unlike anything else in US public land management, the CFLRP is a nationwide program that requires collaboration throughout the life of national forest restoration projects, joining agency partners and local stakeholder groups in a kind of decade-long restoration marriage. Through focused, multi-year investments in particular landscapes, prioritization of funding through a competitive proposal process, and an emphasis on large-scale collaborative planning among state and non-state actors, the approach builds upon existing laws that emphasize synoptic planning and regulation, but adds a policy tool that augments capacity and incentives for groups to organize and work together. The CFLRP was designed to support the places and people that are most poised to be successful in achieving land management goals without compromising the accountability provided by the bedrock of US environmental laws. It also incorporates the flexibility to respond to local needs and contexts. While federal funding investments must be allocated to projects that meet broadly defined goals, the law leaves ample room for local agency personnel and stakeholders to tailor projects to their local social, economic, and ecological conditions.

Consolidating the scholarship on the CFLRP affords us the ability to synthesize what has been learned through the first decade of this program, take stock of successes and challenges, and decide upon next steps. Governance approaches and management goals change as we collectively define and redefine what we want from our national forests and other publicly managed landscapes, and what it means to manage them for ecological, social, and economic sustainability. As we engage in deliberation and action over the generations we want to gather lessons learned and avoid reinvention of the wheel. For any student, practitioner, or policy maker, understanding the central tensions of forest governance and how we have navigated them in the past is critical to charting a better way forward. And, while the CFLRP holds considerable promise, emerging from the community-forestry movement with support from key government actors (see the Foreword), it also requires critical examination. Policies always misbehave, resulting in unexpected outcomes, and the implementation of new policies often reveals a new suite of challenges. Future policy approaches will build on this one, making research on the CFLRP essential for informing future policy and practice in forest management.

Beyond this, as a unique and heretofore untested policy approach, particularly in the US, the CFLRP represents an exciting opportunity for stakeholders, land managers, and researchers to explore the utility of a new type of policy tool for public land governance. After years of demands for greater involvement from community-based stakeholders, the CFLRP presents the most significant pathway for collaborative management on national forests in US public lands management history. In the face of rapid ecological change, the CFLRP also holds potential to address long-standing challenges to support action over sustained time periods, at spatial extents large enough to affect ecological processes without losing sight of the social and economic contexts. As a nationwide policy experiment, the CFLRP provided a unique opportunity for researchers to gather lessons learned across multiple cases on some of the most pressing issues in land management: the dynamics that support successful collaboration; the use of science to improve ecological restoration and adaptive management; and the process of collaborative implementation.

This book brings together chapters from a community of social science researchers who have conducted studies across multiple CFLRP projects to generate insights, not just about the program but also about dynamics that are central to collaborative and large landscape-scale approaches to land management and relevant to broader practice. This introductory chapter describes the historical context for the CFLRP as we consider how forest management goals have changed over the years and explore what this has meant for governance (i.e. the actors and institutions – including laws, agency procedures, and formal and informal decision-making norms – that structure decisions about national forest management). We then devote a section to providing a detailed overview of the CFLRP legislation and its antecedents in collaborative forest restoration policy and practice to set the stage for this book. As scholars of collaborative forest management, we offer insights into some of the most important questions for social scientists studying collaborative forest restoration, and finish with an outline of the organization of the volume.

From scientific management to adaptive governance

The last forty years have witnessed profound changes in US forest governance, with steady movement away from the long-standing foundations of scientific management to today's emphasis on collaborative and adaptive governance (Brunner *et al.*, 2005). These more recent approaches are, in some ways, responses to the proliferation of environmental laws in the 1970s, when the convergence of environmental crises and public advocacy forced policy makers to develop regulatory responses to protect water, air, and biodiversity. This resulted in new legal mandates, like the National Environmental Policy Act of 1969 (NEPA), the Endangered Species Act of 1973 (ESA), and the National Forest Management Act of 1976 (NFMA). Both the NFMA and NEPA were built on the foundation of rational, comprehensive planning and were meant to increase transparency, public involvement, and the use of science in decision-making. While it responded to a demand for greater public participation, the legislation of the era simultaneously deepened the resolve to maintain administrative legitimacy through scientific management principles (Poisner, 1996). By the 1980s, gridlock gripped federal agencies as multiple actors used new legal avenues to challenge agency decisions and force a national reckoning about how to balance commodity production and resource protection. US forest management faced a crisis that garnered national attention, as timber harvests dropped to one tenth of their historic levels by the 1990s. The causes were many, including global economic changes and the legacies of past management, although environmental litigation around endangered species protection and procedural issues associated with NEPA were the most visible suspects (Wear and Murray, 2004). The public narrative of the northwest forest crisis pitted biodiversity and old-growth conservation against forest jobs and rural community well-being. Conflict was high, and, for a time, cooperation seemed unattainable (Layzer, 2009).

Community-based collaborative natural resources management emerged as a way forward. By most accounts, collaboration was initiated by traditional adversaries coming together, during the timber wars in the western United States, to find a path forward that could satisfy both environmental interests as well as local natural resource-dependent economies (Wondolleck and Yaffee, 2000; Weber, 2003). Simultaneously, the emergence of collaborative governance approaches was part of a broader, global shift in environmental governance in the late twentieth century towards 'ecosystem' or 'ecosystem-based' management (Imperial, 1999; Layzer, 2009; Ansell and Gash, 2008). The tenets of ecosystem management, commonly invoked in US forest management in the 1990s, endure in today's goals of restoration and resilience (Predmore *et al.*, 2008; Nie, 2013). Central to ecosystem management were several key ideas: planning for longer-term sustainability goals; focusing on conditions left behind rather than commodities extracted; recognition of the importance of humans, both as components of ecosystems and in defining the sustainability goals for any landscape; the need to work across scales and boundaries; and the need for monitoring and adaptive management to learn in the face of complexity, uncertainty, and change (Grumbine, 1994; Imperial, 1999; Cortner and Moote, 1999).

These goals persist today, and arguably are made more urgent by the current rapidity and extent of ecological change and disturbance (e.g. fire and insect outbreaks). Increasingly, forest management in this century has focused on 'landscape' approaches, which emphasize the need to work across sectors and jurisdictions to achieve multiple resource objectives (Arts *et al.*, 2017). 'Landscape' is a boundary concept, meaning that it is an idea flexible enough for multiple groups of people to coalesce around, without abandoning their distinct goals. In US land management context, landscape approaches involve a strategic plan of action that situates individual management actions or 'treatments' into a broader spatial context, such as that of a watershed or fire shed, with the intention of affecting ecological processes at a meaningful spatial extent.

The rise of collaboration and the need to work across larger landscapes to affect ecological outcomes have changed the nature of forest governance. For one, both require novel arrangements to overcome persistent scalar mismatches (Cumming *et al.*, 2006). Actors must organize for collective action on scales large enough to affect ecological processes and find ways to sustain action and support learning over time. Land managers and their collaborating partners must find ways to work across jurisdictional boundaries and figure out how to support multi-year programs of work, despite the fact that political appropriations and elections cycles, as well as agency promotion cycles, create incentives to focus on short-term, measurable outputs (Biber, 2013).

Public land management agencies, as the primary actors of forest management, face their own distinct struggles. To achieve their goals and respond to change, US federal agencies must identify greater opportunities to network across sectors and with non-state actors – a task that presents a major challenge for US public administration in the twenty-first century (Kettl, 2006). The National Forest System has suffered a precipitous decline in resources and capacity, and the Forest Service today has just over half as many employees in the National Forest System as it did in the year 2000, largely because a growing proportion of its resources have gone to respond to wildland fires (USFS, 2015). These capacity limitations necessitate prioritization of resources in places where the ecological, social, and economic factors align, so that targeted investments are likely to yield successful outcomes. This situation will likely become more acute going forward as governance approaches in the US become increasingly characterized by neoliberal ideals of decreased public investment and greater reliance on private actors to contribute funding and capacity (Abrams *et al.*, 2017).

As we write this Introduction, we find ourselves in a unique and ever-changing era, one that takes us beyond the days of ecosystem management to a movement focused on restoration and resilience, with communities and forests recognized as interdependent facets of an integrated social-ecological system. Stark divisions between science and values, the timber industry and environmentalists, jurisdictional and management boundaries, and the silos of agency bureaucracy are less relevant than they used to be. The dichotomy between preservation and conservation is no longer straightforward, as we recognize, particularly in light of climate change, that no place is left untouched by humans. The dynamics of global change push us away from the past and toward a future of land management that is highly uncertain. Decades of research have shown that sustainability relies on the coupling between human and natural systems, and that while approaches must be informed by national and global institutions and forces, they must be tailored to local contexts. The result of these factors is that environmental policy today must be globally informed, situated within a national framework to provide support and accountability, and community centred in planning what to do and seeing it through on the ground.

In the face of these many challenges, there has been a consistent shift toward collaboration between stakeholders and responsible agency personnel to get the work of public lands management moving again. Collaborative approaches today are more the norm than the exception, both because partners demand a greater role in decision-making and because they bring much needed capacity. The realities of public administration in the US and the nature of contemporary environmental problems necessitate work across jurisdictions and the leveraging of non-state capacity through partnerships and networks. Beyond this, if management is to have any effect on large-scale disturbance processes, actors must find ways to work at an increased pace and scale – a phrase that has been used repeatedly over the last couple of decades. Perhaps what is missing from this phrase, however, is a recognition that our ability to manage landscapes is limited, requiring strategic prioritization and high-quality work – something that may elude easy measurement and messaging. With these many challenges ahead, a central question is how to support these changes through policy and practice.

The foundations of collaborative restoration policy and community-based action

With regard to US national forests, the Forest Service is the primary actor with sole decision-making authority for all actions. In 1960, the US Congress passed a multiple-use mandate (Multiple Use Sustained Yield Act (MUYSA)) for the Forest Service, which, since its inception, had focused primarily on sustainable timber management and watershed protection. By the 1970s, it was clear that, as a result of agency culture, financial incentives, and political pressures, the agency was engaging in a 'conspiracy of optimism', overharvesting timber and failing to make multiple-use a reality (Hirt, 1996; Wilkinson, 1992). In 1976, Congress passed the National Forest Management Act (NFMA), maintaining multiple use as the agency's broad mandate, but putting in place detailed requirements for land and resource management planning (i.e. forest planning), and also setting sideboards to protect resources-at-risk. The NFMA and the many other environmental statutes passed in the 1970s were meant to limit the influence of local interests, political motivations, and professional biases on land management. Collaborative and adaptive governance approaches must be built into the interstitial spaces of this legal-administrative system.

Today, one of the dominant challenges for forest managers is fire, and, through landscape-scale forest restoration, the CFLRP was designed to respond to the challenges of wildfire management. Although fire in the US poses a growing risk to ecosystems and communities, but, because of its natural ecological role, scientists, stakeholders, and policymakers agree there is a need to 'live with fire'. Some forest types may benefit from restoration activities because of over a century of fire exclusion. Restoration activities often involve the reduction of woody biomass (through a process sometimes referred to as mechanical thinning), so that both wildland and prescribed fire, subsequently applied, will pose less risk to resource values and the safety of fire personnel and nearby communities. Where the goal is to influence fire behaviour and reduce fire hazard, efforts must be coordinated across landscapes because of the spatial extent of natural fire events and the contiguity of fuels that drive fire behaviour (North *et al.*, 2012).

Between 2000 and 2005, the average annual acreage burned by wildfire in the United States was 70 per cent greater than in the 1990s (GAO, 2007). Appropriations for responding to wildland fire increased from $1.3 billion on average in the last half of the 1990s, to $3.1 billion on average annually between 2001 and 2005 (numbers adjusted for inflation). By 2015, nearly 60 per cent of the agency's budget was going to fight wildland fire, compromising the Forest Service's ability to achieve restoration to mitigate those very fires before they started, along with every other aspect of its mission (USFS, 2015). In the face of these challenges, states and communities had been working together to build agreement and capacity to address the threat of fire and support restoration. As just one example, in Arizona, Governor Napolitano convened a Governor's Forest Health Council, which developed a Statewide Strategy for Restoring Arizona's Forests (Governor's Forest Health Councils State of Arizona, 2007). In northern Arizona, with financial support from the Forest Service, community members developed an assessment of small-diameter wood supply study, in order to identify opportunities for new industries to locate in the area and complete much-needed restoration work in a way that would have broad social support. These efforts paved the way for the CFLRP project in that region (Hampton *et al.*, 2008).

Other groups across the nation had participated in a joint venture started in 2001 by federal land management agencies and The Nature Conservancy, called the US Fire Learning Network, a peer-learning network that engaged participants on regional and national scales in sharing strategies for addressing landscape-scale fire management challenges (Butler and Goldstein, 2010). Participants in the network developed large-landscape and cross-boundary restoration plans for fire adapted ecosystems to influence state, federal, and private land management strategies across a region. Staff from the Nature Conservancy and participants from the Fire Learning Network recognized that restoration plans needed dedicated funding if they were going to be implemented, and worked to develop and support the legislation that established the CFLRP (Butler and Goldstein, 2010). When the CFLRP became a reality, three of the funded projects had been participants in the Fire Learning Network and had prepared themselves for the implementation phase of ecological fire restoration projects.

Leading up to the CFLRP, the community-based forestry movement had identified several changes needed to support collaborative restoration, including: the need for long-term and flexible funding to allow for multi-resource restoration activities over a large enough landscape to affect ecological processes; identification of innovative contracting mechanisms; support for collaboration with a focus on monitoring and performance indicators to improve accountability; and a continued need to improve and make NEPA processes more efficient (Cromley, 2005). Policy changes followed, often due to mobilization across community-based forestry leaders, working together with key policy makers and agency leaders. Stewardship contracts allow the agency to bundle goods, such as marketable forest products, with services, such as the removal of restoration by-products, into a single contract. This type of contracting can be useful for accomplishing restoration and supporting local economies, particularly when certain forest products are of little value while new markets are being created. According to the General Accountability Office (GAO), 'The agencies credit stewardship contracting with allowing them to accomplish more work … and spurring collaboration with members of the community and environmental groups' (GAO, 2008). Between 2013 and 2016, the Integrated Resource Restoration budget pilot, another rallying point for the Rural Voices for Conservation Coalition (RVCC), combined previously separate budget lines to allow for greater budgetary flexibility. Although the pilot ended, the Forest Service continued to utilize tenets of the approach, and supported a more flexible use of appropriations. These budget and contracting approaches facilitated implementation during the CFLRP.

Other policy approaches to address the fire crisis and support restoration also yielded important lessons that influenced the CFLRP. The CFLRP legislation, championed by the Senate Energy and Natural Resources chairman, Jeff Bingaman (D-NM), was based in part on the successes and lessons learned under the Community Forest Restoration Act of 2000, which offered cost-share grants for collaboratively designed forest restoration projects in Bingaman's home state of New Mexico. Congress also addressed fire management with the passage of the Healthy Forests Restoration Act (HFRA) in 2003, which involved communities in defining fuels treatment priorities through Community Wildfire Protection Plans. However, as fire suppression costs increased, money that had been allocated to community-supported fuels reduction projects under the HFRA was redirected to pay for suppression, and never replenished (Cromley, 2005). In addition, the HFRA did little to reduce conflict, incorporate collaborative agreement into project design, or prioritize landscapes for treatment (US Senate Committee on Energy and Natural Resources, 2006). This created an opportunity for a new policy that would address these issues to gain support from Congress and community-based forestry advocates.

A central question, then, was how to design a policy that could support collaboratively forged agreements, without being overly specific or alienating agency partners. Previous place-based legislation had not been successful. One of the most famous early natural resource based collaborative efforts was the Quincy Library Group, which came into being when environmentalists and members of the timber industry sought to break what they perceived to be an impasse in forest management in the Sierra Nevada. In the face of agency reluctance to convert their agreements into plans, the group took their cause to Congress, which passed the Herger-Feinstein Quincy Library Group Forest Recovery Act of 1998. However, because the Forest Service retained decision-making authority and faced overlapping and conflicting legal requirements, the agency failed to honour the agreements reached by the

Quincy Library Group, even when these had been mandated through place-based legislation. Specific requirements of the law also became less applicable to the context over time, as landscapes and markets changed (Cheng *et al.*, 2016). Other attempts at place-based legislation were made, but without much success (Nie and Fiebig, 2010). The lesson was that, given legal complexity and changing conditions, both agency buy-in and some flexibility would be needed to support the implementation of collaborative agreements.

Design of the CFLRP

The Forest Landscape Restoration Act (PL. 11–111; 16 USC §7303) that established the CFLRP was introduced by Senator Bingaman in 2008 and passed in 2009 as part of a broader package of public lands legislation (for more on this story, see the Foreword). The stated purpose of the Act was to 'encourage the collaborative, science-based ecosystem restoration of priority forest landscapes' (16 USC §7301). Hearings leading up to its passage emphasized the need to address wildland fire and associated costs and threats to forest ecosystems, and to move beyond incremental approaches to forest restoration. By emphasizing collaboration, economic feasibility, and planning across large landscape areas, the CFLRP offered a way to address the need for forest restoration on larger scales than had been feasible under past programs and legislation.

The language of the Act lays out a number of key objectives: promoting ecological, economic, and social sustainability; leveraging local resources to accomplish these goals; reducing fire management costs through reducing fire hazard and restoring natural fire regimes; demonstrating the degree to which restoration activities achieve ecological/watershed objectives and affect fire activity and its associated costs; and showing how capturing the value of forest restoration by-products could reduce treatment costs and support local economies. The law requires projects to be based on a restoration strategy that prioritizes treatments for a ten year period across landscapes of at least 50,000 acres in size, comprised primarily of National Forest System lands. Projects cannot build permanent roads or remove old-growth, and are supposed to promote habitat and watershed improvement, support the application of prescribed fire and the return of natural fire, and include measures to control invasive species. All projects have to be 'developed and implemented through a collaborative process' (16 USC §7303(b)(2)) and have to explain how they will use existing or proposed industry infrastructure to process restoration by-products in order to support jobs, local economic development, and achieve a reduction of costs associated with restoration treatments. All projects must also undertake multi-party monitoring of social, ecological, and economic outcomes for at least fifteen years to support learning and ensure accountability over time. The budget line item for CFLRP allows the Forest Service the flexibility to use funds to support any of these objectives; however, implementation can take place only on the National Forest System lands.

The law creates a competitive funding program, soliciting proposals developed jointly by the US Forest Service and stakeholders for landscape-scale restoration programs of work. When it passed, it was wholly unique, both in establishing competition for funding and in supporting prioritization of investments based on proposals that justified the investment due to a confluence of social agreement, ecological need, and economic viability for restoration. Congress authorized up to $40 million annually for competitive allocation under the program and required an application process, with selections made by the Secretary of Agriculture based on input from a Federal Advisory Committee chartered expressly for this purpose. In 2010, ten projects were selected. These projects received the $10 million appropriated in fiscal year 2010, and another $25 million in appropriations in 2011. In 2012, the full $40 million was appropriated, which supported the original projects and ten new projects. The Forest Service also contributed other appropriations to add three more projects, bringing the total to twenty-three projects, ten from 2010, and thirteen from 2012 to 2013. The authority for the CFLRP does not expire, although in 2019 the authority for appropriations requires renewal from Congress.

Research priorities for collaborative governance and the CFLRP

After nearly four decades of research on collaborative governance and particularly collaborative environmental management, what remains for us to learn? The CFLRP policy experiment allowed social scientists to examine questions that have yet to be resolved and in a particularly interesting context. There are several features that attracted the scholars included in this volume to study CFLRP cases. First, the collaborative governance literature is dominated by case studies of relatively limited time periods – snapshots of collaborative processes (Ansell and Gash, 2008; Margerum and Robinson, 2017). With the commitment to fund CFLRP projects for ten years, these cases offer a rare opportunity for longitudinal study and analysis. Second, CFLRP cases offer an equally rare opportunity to conduct a comparative analysis of multiple collaborative groups with distinct characteristics, but all operating under the same policy context. Collaborative processes emerge under unique conditions and are attuned to the peculiarities of the context in which they arise. Policies that mandate collaboration are increasingly prevalent in natural resources and environmental management (Cheng, 2006; Monroe and Butler, 2015), and there is much more to learn from these types of policy experiments, particularly with multiple sites of implementation. Third, much of the collaborative environmental management literature consists of cases of collaborative participants coming to agreement on specific plans of action, but rarely continue beyond that point (Margerum, 2011; Koontz and Newig, 2014). Because the legislation required CFLRP projects to engage in collaboration with stakeholders throughout planning, implementation and monitoring, this program offered a unique window into how collaboration can shape implementation and expand accountability of government agencies through multi-party monitoring. Finally, the CFLRP legislation implies that adaptive management is a central goal of the program. Adaptive management has been elusive in practice, despite several decades of scientists and managers advocating its logic (Lee, 1999; Biber, 2009; DeLuca *et al.*, 2010; Ruhl and Fischman, 2010). The CFLRP provides a venue within which adaptive management experiments can be tried and examined.

All of these facets of CFLRP promise both practical and theoretical insights. Practically, we are interested in several key questions. To what extent and how has it achieved the goals of improved collaboration; increased the pace, scale and quality of ecological restoration; led to improved monitoring and adaptive management; and struck the delicate balance between higher levels of stability from the federal government with the flexibility to tailor approaches to local contexts? Theoretically, this program pushes scholars to refine our understanding of collaborative governance processes, their design, effectiveness, and contributions to public lands management where authority and responsibility is vested within land management agencies. While the role of government agencies has long been understood as an essential element of collaborative governance of public issues (Ansell and Gash, 2008; Wondolleck and Yaffee, 2000; Koontz *et al.*, 2004), the promise of collaboration is that it expands the democratic space to involve governance *with* government (Agranoff, 2008; Weber and Khademian, 2008). Public lands management is a particularly interesting context due to the longstanding legal and regulatory framework that has legitimized a scientific management process and placed the ultimate authority and responsibility for decision-making and action in the hands of public agencies. Thus, examining how collaboration functions within such a context has the potential to provide insights that are applicable to a wide range of natural resources management contexts. What works in this context is likely to function in contexts where the authority and responsibility is less clearly delineated. The authors in this volume address questions under three thematic areas that pertain to this context: collaborative governance dynamics; science and adaptive management; and collaborative implementation.

Organization of the book and chapter overviews

This book is written to share the compendium of social science research on the CFLRP and stimulate dialogue among students, scholars, practitioners, and policy makers. Each chapter begins with an introduction to relevant theory and concludes with policy and practice implications. The book proceeds in three main sections, and each benefits from a commentary by individuals who are both scholars and practitioners working in collaborative environmental management contexts. The commentaries provide insight into the key contributions and future directions in policy and scholarship that this work portends. The first section examines questions about collaborative governance processes and the dynamics of trust, accountability, and capacity. The second section clarifies how scientific information is used in making decisions, coming to consensus, and contributing to adaptive management through the collaborative process. Finally, the third section examines the topic of collaboration through implementation, an underdeveloped area of collaborative governance literature.

Section 1: collaborative governance dynamics

In the context of landscape-scale management, it is critical to understand collaborative group characteristics that support success. The first section focuses on collaborative dynamics of trust, accountability, and capacity. While these themes have been identified as important topics in collaborative environmental management literature for several decades, they have not been sufficiently addressed in prior research (Wondolleck and Yaffee, 2000; Emerson and Nabatchi, 2015; Margerum and Robinson, 2017). The first chapter by Stern and Coleman discusses the critical role of trust for enduring collaboration and provides strategies for assessing whether and how collaboratives are organized to enhance trust building. Trust has long been conceived of as an essential ingredient for successful collaboration, and a proxy for the quality of relationships in collaborative groups. However, Stern and Coleman clarify that trust is not a monolithic concept. They elaborate on their own typology of trust based on social-psychological and natural resources literature (Stern and Coleman, 2018) and apply these concepts to CFLRP cases, which they used to build a theory of trust ecology. The next chapter, by Christensen and Butler, identifies potential tensions that arise when collaborative partners and agency staff operate under overlapping and sometimes competing accountability systems. It then enumerates strategies that CFLRP participants use to soften the tensions between organizational requirements and the need to maintain accountability between the agency and stakeholders. Accountability in representative democratic government follows relatively clear pathways through legislative, administrative, and judicial processes. Collaborative governance operates alongside this formal government system, complicating the accountability context (Wondolleck and Yaffee, 2000; Weber, 2003; Bryson *et al.*, 2015). Christensen and Butler seek to demystify this context to explore how multiple forms and manifestations of accountability interact between the formal government system and the less formal collaborative system, and, importantly, what activities seem to bridge these two domains to ensure effective accountability between government agencies and collaborative groups. The last chapter, by Ryan and Urgenson, explores how collaborative groups are able to build capacity to develop shared visions of restoration. The collaborative capacity building literature points to a variety of factors that enable collaborative groups to function effectively. The authors elaborate on how specific facets of the CFLRP cases they studied contributed to enhancing collaborative capacity and what challenges remain where capacity gaps exist.

Section 2: science and adaptive management in collaborative restoration

The second section of the book consists of three chapters that focus on the challenges of operationalizing broad scientific concepts in practice with collaborative partners where social and ecological considerations interact. This involves identifying relevant science, monitoring strategies and their implications for adaptive management, and specific land management goals. Toman, Walpole, and Heeren address how practitioner mental models shape visions for how forest restoration and resilience can be achieved. Based on the mental models of study participants, the authors find that restoration as a concept can galvanize collaborative groups. However, restoration is also a contested idea. While collaboratively developing a restoration vision is important to guide planning and implementation and influence agencies, it can also leave some challenges unresolved. The second chapter by Cheng, Aplet, and Waltz explicates how three CFLRP projects took adaptive management concepts and integrated them into the practice of collaborative landscape scale restoration. The authors elaborate on the processes and factors that facilitate and frustrate practitioners in achieving the goals of adaptive management and what CFLRP collaborative groups did to overcome some of the barriers. In the final chapter of the section, Colavito describes how science is identified and used to inform decisions within collaborative contexts, which carries implications for how science and policy interact in on-the-ground restoration action. In particular, she examines the role that boundary organizations can play in interpreting and communicating science to collaborative groups and agency personnel, generating new information through monitoring, and resolving conflicts about contested science on landscape projects. This section addresses the disconnect between adaptive management in theory and practice by providing practical strategies to more effectively undertake adaptive management in forest restoration and address the linkages between science, information, values, and restoration activities.

Section 3: collaborative implementation

The final section synthesizes the key insights related to how, why, and to what end collaborators work with agency partners throughout the implementation and monitoring of landscape-scale projects. Butler, Monroe, and McCaffrey explore the implications of collaborative implementation in the adaptive management process to influence agency actions over time. They find that the implementation process is highly connected to planning and monitoring for adaptive management purposes. Moreover, given the largely indirect influence that collaborative groups inherently have on agency actions, the role of collaborative groups in monitoring and ongoing prioritization are highly valuable contributions during the implementation phase. The authors conceptualize an adaptive management process that operates on multiple temporal and spatial scales while also connecting this process to the increased pace, scale, and quality of restoration treatments. The second chapter by Bergemann, Schultz, and Cheng aims to clarify how the role of past experiences in collaboration can influence the level and quality of participation in collaborative implementation. Through comparative case analysis, they find that the role of history and a legacy of trust, the posture of local implementers vis-à-vis the collaborative group, and the ongoing commitment of leaders to interact with collaborative groups create collaborative characteristics specific to each landscape that shape how implementation unfolds. Finally, Schultz and McIntyre undertake a comprehensive evaluation of the CFLRP to investigate what benefits were realized under the CFLRP, what factors most influenced project success, and what improvements may be needed in future policy approaches. Based on a survey of agency personnel and more than eighty interviews in year eight of the CFLRP, they find that the focused, multi-year investment, and requirements to collaborate through implementation and monitoring supported improved relationships, greater collective action, larger-scale planning, and innovations in multiple areas.

Concluding thoughts

The book concludes with a synthesis of ideas that contribute to scholarship of collaborative environmental management, the policy and practice of large scale collaborative forest restoration on public lands, and a series of future research questions. These ideas emerged in part from the chapters herein and in part through a symposium that brought together the authors included in this volume, along with practitioners from key organizations who have participated in CFLRP since its inception. Through the analysis of the program and the projects within it, we contribute to the theory of collaborative environmental management and to a future research agenda.

The CFLRP portends a sea change in forest management. To see the transformation come to full fruition, the agency, collaborators, and Congress will have to commit to further investments of time, energy, creativity, and funding. We envision a future where collaborative and adaptive governance is the norm, where restoration means the integration of social and ecological system issues in the face of an uncertain future, and where creative solutions to longstanding problems of forest governance can be attempted to enhance learning and adaptation. In the final chapter, we lay out not only what this vision entails, but how to get there through changes to practice, policy, and administration.

While our observations and recommendations represent important contributions to the study of collaborative governance, they cannot substitute for the hard work of those who engage in its practice. As discontent with the federal government continues to grow with some segments of the public, the best opportunity to resolve natural resource conflicts resides with local people working together at to solve local problems. This was the story in Harney County, and the story we have observed over the course of collective work on the CFLRP, a program that has fostered a spirit of cooperation and collaboration to address vexing land management problems. This volume seeks to capture the lessons learned from this experiment to inform the future of US land management in particular, but also inspire and enhance collaborative forest restoration activities globally.

References

Abrams, J.B., Huber-Stearns, H.R., Bone, C., Grummon, C.A., and Moseley, C. (2017) 'Adaptation to a landscape-scale mountain pine beetle epidemic in the era of networked governance: The enduring importance of bureaucratic institutions', *Ecology and Society*, 22(4)

Agranoff, R. (2008) 'Enhancing performance through public sector networks: Mobilizing human capital in communities of practice', *Public Performance & Management Review*, 31(3): 320–347

Ansell, C. and Gash, A. (2008) 'Collaborative governance in theory and practice', *Journal of Public Administration Research and Theory*, 18(4): 543–571

Arts, B., Buizer, M., Horlings, L., Ingram, V., Van Oosten, C., and Opdam, P. (2017) 'Landscape approaches: A state-of-the-art review', *Annual Review of Environment and Resources*, 42(1): 439–463

Biber, E. (2009) 'Too many things to do: How to deal with the dysfunctions of multiple-goal agencies', *Harvard Environmental Law Review*, 33: 1–63

Biber, E. (2013) 'Adaptive management and the future of environmental law', *Akron Law Review*, 46(4)

Brunner, R.D., Steelman, T.A., Coe-Juell, L., Cromley, C.M., Edwards, C.M., and Tucker, D.W. (2005) *Adaptive Governance: Integrating Science, Policy, and Decision Making*. New York: Columbia University Press

Bryson, J., Crosby, B., and Stone, M. (2015) 'Designing and Implementing Cross-Sector Collaborations: Needed and Challenging', *Public Administration Review*, 75(5): 647–663

Butler, W.H. and Goldstein, B.E. (2010) 'The US Fire Learning Network: Springing a rigidity trap through multiscalar collaborative networks', *Ecology & Society*, 15(3)

Cheng, A.S. (2006) 'Build it and they will come? Mandating collaboration in public lands planning and management', *Natural Resource Journal*, 46(4): 841–858

Cheng, A.S., Gutiérrez, R.J., Cashen, S., Becker, D.R., Gunn, J., Merrill, A., Ganz, D., Liquori, M., Saah, D.S., and Price, W. (2016) 'Is there a place for legislating place-based collaborative forestry proposals?: Examining the Herger-Feinstein Quincy Library Group Forest Recovery Act Pilot Project', *Journal of Forestry*, 114(4): 494–504

Cortner, H. and Moote, M.A. (1999) *The Politics of Ecosystem Management*. Washington DC: Island Press

Cromley, C.M. (2005) 'Community-based forestry goes to Washington'. In R.D. Brunner, T.A. Steelman, L. Coe-Juell, C.M. Cromley, C.M. Edwards, and D.W.

Tucker (Eds.) *Adaptive Governance: Integrating Science, Policy, and Decision Making.* New York: Columbia University Press

Cumming, G.S., Cumming, D.H.M., and Redman, C.L. (2006) 'Scale mismatches in social-ecological systems: Causes, consequences, and solutions', *Ecology and Society*, 11(1)

DeLuca, T.H., Aplet, G.H., Wilmer, B., and Burchfield, J. (2010) 'The unknown trajectory of forest restoration: A call for ecosystem monitoring', *Journal of Forestry*, 108(6): 288–295

Emerson, K. and Nabatchi T. (2015) *Collaborative Governance Regimes.* Washington DC: Georgetown University Press

Grumbine, R.E. (1994) 'What is ecosystem management?', *Conservation Biology*, 8(1): 27–38

Governor's Forest Health Councils State of Arizona. (2007) *The Statewide Strategy for Restoring Arizona's Forests.* In E. Aumack, T. Sisk, and J. Palumbo (Eds). Arizona Public Service Company, Phoenix

Hampton, H.M., Sesnie, S.E., Dickson, B.G., Rundall, J.M., Sisk, T.D., Snider, G.B., and Bailey, J.D. (2008) Analysis of Small-Diameter Wood Supply in Northern Arizona. Forest Ecosystem Restoration Analysis Project, Center for Environmental Sciences and Education. Northern Arizona University, Flagstaff

Hirt, P.W. (1996) *A Conspiracy of Optimism: Management of the National Forests since World War Two.* Lincoln, Nebraska: University of Nebraska Press

Imperial, M.T. (1999) 'Institutional analysis and ecosystem-based management: The institutional analysis and development framework', *Environmental Management* 24(4): 449–465

Kettl, D.F. (2006) 'Managing boundaries in American administration: The collaboration imperative', *Public Administration Review*, 66(1): 10–19

Koontz, T.M. and Newig, J. (2014) 'From Planning to Implementation: Top-Down and Bottom-Up Approaches for Collaborative Watershed Management', *Policy Studies Journal*, 42: 416–442

Koontz, T., Steelman, T., Carmin, J., Korfmarcher, K., Moseley, C., and Thomas, C. (2004) *Collaborative Environmental Management: What Roles for Government?* Washington DC: Island Press

Layzer, J.A. (2009) *Natural Experiments: Ecosystem-based Management and the Environment.* Cambridge, MA: MIT Press

Lee, K.N. (1999) 'Appraising adaptive management', *Conservation Ecology*, 3(2)

Margerum, R.D. (2008) 'A typology of collaboration efforts in environmental management', *Environmental Management*, 41(4): 487–500

Margerum, R. (2011) *Beyond Consensus: Improving Collaborative Planning and Management.* Cambridge, MA: MIT Press

Margerum, R.D. and Robinson, C.J. (2017) *The Challenges of Collaboration in Environmental Governance: Barriers and Responses.* Cheltenham: Edward Elgar Publishing

Monroe, A.S. and Butler, W. (2015) 'Responding to a policy mandate to collaborate: Structuring collaboration in the Collaborative Forest Landscape Restoration Program', *Journal of Environmental Planning and Management*, 59(6): 1054–1072

Nie, M. (2013) 'Whatever Happened to Ecosystem Management and Federal Lands Planning?' In K. Robbins (Ed.) *The Laws of Nature: Reflections on the Evolution of Ecosystem Management Law and Policy.* Akron, OH: University of Akron Press

Nie, M. and Fiebig, M. (2010) '*Managing the national forests through place-based legislation*', *Ecology Law Quarterly*, 37(1): 1–52

North, M., Collins, B.M. and Stephens, S. (2012) 'Using fire to increase the scale, benefits, and future maintenance of fuels treatments', *Journal of Forestry*, 110(7): 392–401

Poisner, J. (1996) 'A civic republican perspective on the National Environmental Policy Act's process for citizen participation', *Environmental Law*, 26(1): pp. 53–94

Predmore, S.A., Copenheaver, C.A., and Mortimer, M.J. (2008) 'Ecosystem management in the US Forest Service: A persistent but dying discourse', *Journal of Forestry*, 106(6): 339–345

Ruhl, J.B. and Fischman, R.L. (2010) 'Adaptive management in the courts', *Minnesota Law Review*, 95(2): pp. 424–484

Stern, M.J. and Coleman, K.M.J. (2018) 'Exploring the functions of different forms of trust in collaborative natural resource management', *Society & Natural Resources*, 31(1): 21–38

US Forest Service (USFS). (2015) *The Rising Cost of Wildfire Operations: Effects on the Forest Service's Non-Fire Work*. US Department of Agriculture, USDA Forest Service

US Government Accountability Office (GAO). (2007). *A cohesive strategy and clear cost-containment goals are needed for federal agencies to manage wildland fire activities effectively*. GAO-07–1017T, Testimony before the Subcommittee on National Parks, Forests and Public Lands, Committee on Natural Resources, House of Representatives. US GAO, Washington DC

US Government Accountability Office (GAO). (2008). *Use of stewardship contracting is increasing, but agencies could benefit from better data and contracting strategies*. GAO-09-23, US GAO, Washington DC

US Senate Committee on Energy and Natural Resources. (2006) Implementation of the Healthy Forests Restoration Act. 109th Cong., 2d sess., S. Hrg. 109–506

Wear, D.N. and Murray, B.C. (2004) 'Federal timber restrictions, interregional spillovers, and the impact on US softwood markets', *Journal of Environmental Economics and Management*, 47: 307–330

Weber, E.P. (2003) *Bringing Society Back In: Grassroots Ecosystem Management, Accountability, and Sustainable Communities*. Cambridge, MA: MIT Press

Weber, E.P. and Khademian, A. (2008) 'Wicked problems, knowledge challenges, and collaborative capacity builders in network settings', *Public Administration Review*, 68(2): 334–349

Wilkinson, C. (1992) *Crossing the Next Meridian: Land, Water, and the Future of the West*. Washington DC: Island Press

Wondolleck, J.M. and Yaffee, S.L. (2000) *Making Collaboration Work: Lessons from Innovation in Natural Resource Management*. Washington DC: Island Press

1 Collaborative forest landscape restoration in action

An overview of the CFLRP cases

William H. Butler and Bryce Esch

Introduction

This chapter provides a brief overview of the twenty-three landscape projects and collaborative groups participating in the Collaborative Forest Landscape Restoration Program (CFLRP). The purpose of the chapter is to introduce the range of projects in the program and describe a few key characteristics of each case, so that in subsequent chapters authors can refer to the cases in broad terms and move directly into their analysis of each case relative to the conceptual and analytical frameworks of their respective chapter. We describe location, ecological context, general objectives, landscape size, ownership, and basic collaborative characteristics of the CFLRP cases in general terms. We present more specific data on each case in tables. When reading chapters later in the book, readers may find that referring to this chapter helps situate the cases in the broader program, while providing some details that clarify attributes of each case more specifically.

The authors of this chapter each completed independent research projects on CFLRP landscape cases. We draw on data gathered in these two research projects to provide background on the twenty-three CFLRP cases. With the support of the US Forest Service Northern Research Station and The Nature Conservancy's US Fire Learning Network, Butler spent six years studying collaborative groups from the first ten CFLRP landscapes, focusing on how they transitioned from collaborative planning to implementation. During that project, he and his research team completed more than 120 interviews; conducted site visits and observations on most of the landscapes and at regional and national meetings; and examined documents, websites, reports, meeting minutes, and other products of the CFLRP landscape project teams and collaborative groups. The research included reviewing landscape project proposals and gathering data on the characteristics of each landscape collaborative group, which informs the descriptions of each of the cases from the 2010 funded projects.

As part of her masters degree capstone project under the direction of Steven Yaffee, and with the support of the National Forest Foundation, Esch completed a one-year assessment of the second round of CFLRP landscape collaborative groups with a team of masters students in the Natural Resources program at the University of Michigan. Throughout 2013, the team conducted sixty interviews and twelve site visits, collecting information on the characteristics, challenges, and successes for each restoration project. They also reviewed the project proposals to gather data on key project characteristics. This data was used in a cross-case analysis of commonalities and differences among the 2012 and 2013 projects, and in a report summarizing case characteristics and providing recommendations for future programs (Antuma *et al.*, 2014). These data and analysis efforts inform our description of the thirteen projects funded in years 2012 and 2013.

Landscape project characteristics

One way to distinguish CFLRP landscape projects is through the basic characteristics of the biophysical, institutional, and social aspects of the projects themselves. In this section, we provide a general overview of the location, ecological systems, size of landscapes, proposed treatment area, NEPA-approved acres, ownership and management boundaries, and funds requested. We primarily report these general characteristics from the project proposal documents with a few updates from subsequent project-level case study research and a review by Washington Office staff. We provide specific data on each project in Table 1.1.

Location

The CFLRP landscape projects cover national forests in nearly all of the US Forest Service regions, with a preponderance of sites in the western United States. The only landscape project east of the Mississippi funded in 2010 was the Accelerating Longleaf Pine Restoration Project in the Osceola National Forest in Florida. Other sites were in Washington, Oregon, Colorado, Arizona, New Mexico, California, Idaho, and Montana. In the round of landscapes selected in 2012/13, sites in the southeast were added in Mississippi, Arkansas, Missouri, and North Carolina although western states continued to be focal with two new sites added in each of Idaho, Oregon, and California, and individual sites added to Washington and New Mexico. As of 2018, there are thirty-four National Forests enrolled in the program, which covered fourteen states. The states with the highest number of projects were Idaho, Oregon, and California, with three landscape projects in each state. The state with the largest land area covered by CFLRP projects was Idaho with nearly three million acres, while the largest project was the Four Forest Restoration Initiative (4FRI) with more than 2.4 million acres in Arizona.

Ecosystems and restoration objectives

Projects vary in their restoration objectives, but all are implementing actions to restore the structure and composition of their landscape ecosystems. Some

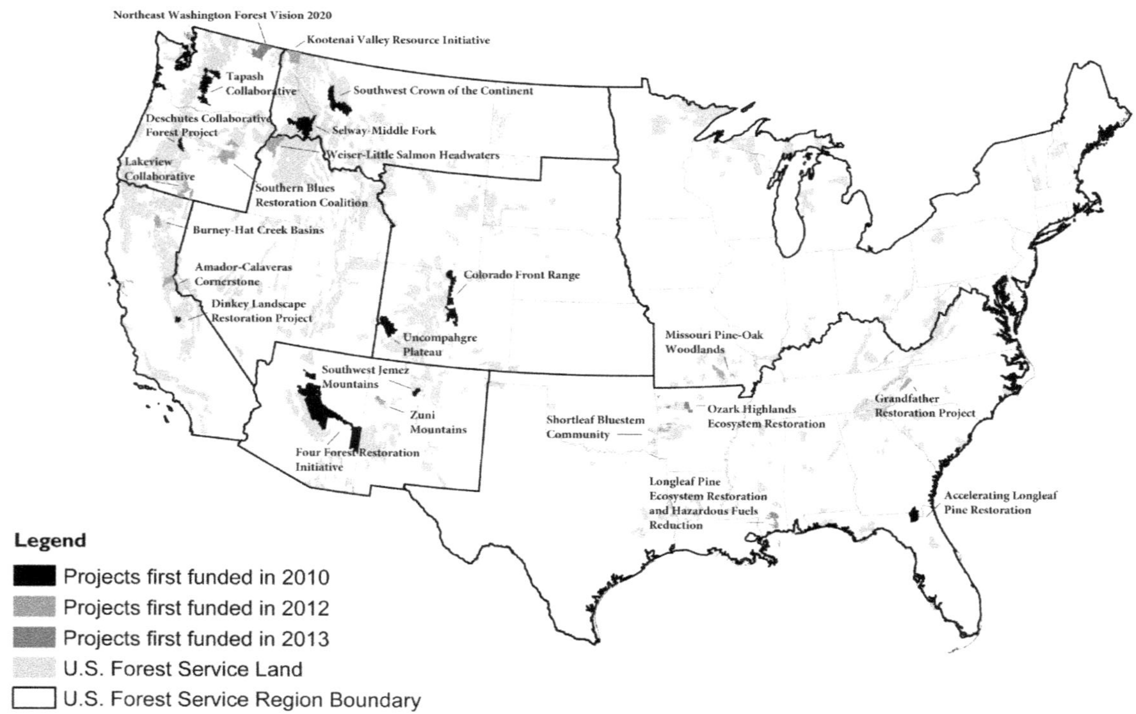

Figure 1.1 Map of CFLRP landscapes.

Source: United States Forest Service Forest Management (public domain).

landscapes, such as the Four Forests Restoration Initiative (4FRI) in Arizona, essentially have one dominant vegetation type (Ponderosa Pine in that case). Others have greater variability across the landscape, such as the Deschutes in Oregon which has four distinct forest types (ponderosa pine, wet mixed conifer, dry mixed conifer, and hemlock). Many projects aim to restore a specific ecosystem type, such as the Shortleaf Bluestem, Ozark Highlands, and Pine-Oak Woodlands projects, all of which propose to increase the shortleaf-pine-oak mosaic ecosystem. These projects fall across the Ouachita and Ozark mountains in the southeastern United States, an area identified by The Nature Conservancy as a biodiversity hotspot due to its unique ecosystem (The Nature Conservancy, 2003). Like many of the other landscapes, much of that ecosystem has been significantly altered through past management practices. Some projects focus on species composition and increasing habitat for target species of interest which are generally adapted to a fire prone landscape. For example, the Longleaf projects in Florida and Missouri both aim to restore Longleaf pine ecosystems to expand habitat for a series of endemic, endangered, and keystone species including the red-cockaded woodpecker (*Leuconotopicus borealis*) and gopher tortoise (*Gopherus polyphemus*). The Longleaf Pine ecosystem and wildlife within it are fire dependent; thus, restoring a natural fire regime is an underlying restoration goal.

While the ecosystem types vary, all projects have fire-adapted or fire-dependent vegetation on part or all of the landscape area covered. Reducing wildland fire risks and management costs was one of the key motivations for establishing the CFLRP (see the Introduction and Foreword in this volume). Western sites are more focused on reducing the immediate risk of severe and large-scale wildfires. Massive wildfires in the west have become one of the most pressing issues addressed through restoration, as fires have increased in magnitude and severity and fire seasons have become longer (Westerling *et al.*, 2006). Most projects proposed to restore forests that had become dense stands of similarly sized and aged trees due to fire exclusion and past timber management practices (Covington and Moore, 1994; Hessburg *et al.*, 2005). These homogenous landscapes can be highly vulnerable to disturbance from fires and the expected effects of climate change (Buma and Wessman, 2013; Keane *et al.*, 2002). This has motivated many groups to come together to solve the fire crisis. For example, collaboration in the Southwest Jemez Mountains emerged in the ashes of a series of large wildfires around the turn of the twentieth century. Fire events in Oregon, Washington, and Colorado spurred the formation of collaborative groups focused on fire management and ecological restoration in those states as well. Most projects proposed to restore forests that had become dense stands of similarly sized and aged trees. This involves reducing tree densities through mechanical thinning and using prescribed fire to mimic natural fire regimes.

Table 1.1 Summary landscape project characteristics

Project name	*State*	*USFS region*	*National forest(s)*	*Landscape area (acres)*	*Proposed treatment area (acres)*	*NEPA acres approved at time of selection*	*CFLR funds requested*
2010 CFLRP Projects							
Selway-Middle Fork Clearwater Project	ID	1	Nez Perce, Clearwater and Bitterroot	1,400,000	52,700	~ 7,400	$36,862,240
Southwestern Crown of the Continent	MT	1	Lolo, Flathead, and Helena	1,449,670	199,140	34,300	$37,029,292
Colorado Front Range Landscape Restoration Initiative	CO	2	Arapaho and Roosevelt, Pike and San Isabel	~800,000	31,600	100,000	$37,000,000
Uncompahgre Plateau Collaborative Restoration Project	CO	2	Grand Mesa, Uncompahgre, and Gunnison	1,000,000	572,000 (160K on USFS lands)	86,700	$8,588,900
Four Forests Restoration Initiative	AZ	3	Apache-Sitgreaves, Coconino, Kaibab, and Tonto	~2,400,000	600,000+ (up to whole project area)	~83,000	$38,000,000
Southwest Jemez Mountains	NM	3	Santa Fe NF and Valles Caldera National Preserve[1]	210,000 (123,146 USFS after transfer)	156,403 USFS; 10,400 non-USFS	35,900	$34,276,000
Dinkey Landscape Restoration Project	CA	5	Sierra	154,000	95,000	~7,000	$5,266,109[2]
Deschutes Collaborative Forest Project[3]	OR	6	Deschutes	130,000 (revised to 257,850)	60,000	38,400	$10,057,000
Tapash Sustainable Forest Collaborative	WA	6	Okanogan-Wenatchee	1,629,959	168,617	58,405	$12,011,028[4]
Accelerating Longleaf Pine Restoration in NE FL	FL	8	Osceola	567,800	unclear	fire management activities; 3 years of thinning	$17,181,682
2012/2013 CFLRP Projects[5]							
Kootenai Valley Resource Initiative	ID	1	Idaho Panhandle National Forests	800,000	39,430	8,150	$12,272,443
Weiser-Little Salmon Headwaters Project	ID	1	Payette	798,900	190,000	50,000	$37,400,000

Zuni Mountain	NM	3	Cibola	210,000	56,000	24,000	$7,600,000
Amador-Calaveras Consensus Group Cornerstone Project	CA	5	Stanislaus, Eldorado	390,904	38,500	1,080	$16,653,850
Burney-Hat Creek Basins	CA	5	Lassen	369,036	69,239	42,400	$10,782,680
Southern Blues Restoration Coalition	OR	6	Malheur	690,723 (revised to 877,900)	271,980	113,097	$22,500,000 (revised to $27,500,000)
Lakeview Stewardship Project	OR	6	Fremont-Winema	662,289	150,000	49,000	$28,100,000
NE Washington Forest Vision 2020	WA	6	Coleville	916,283	124,396	49,206	$31,753,928
Grandfather Restoration Project	NC	8	Pisgah	330,360	41,685	8,954	$4,547,622
Longleaf Pine Ecosystem and Hazardous Fuels Reduction	MS	8	DeSoto	382,000	374,000	382,000	$25,321,024
Ozark Highlands Ecosystem Restoration	AR	8	Ozark and Saint Francis	344,393 (revised to 528,448)	217,892	162,743	$15,808,746
Shortleaf Bluestem Community Project	AR/ OK	8	Ouachita	348,482	320,000	256,447	$21,927,000
Missouri Pine Oak Woodlands Restoration	MO	9	Mark Twain	345,710	115,860	26,000	$12,526,632

Notes
1 The 2014 National Defense Authorization Act transferred the 88,900 acre Valles Caldera National Preserve to the National Parks Service.
2 Included funding for FY2020 in proposal– FY 2010 to FY 2019 total from proposal is $13,597,597.
3 Changed: the collaborative name to Deschutes Collaborative Forest Project (DCFP) subsequent to proposal as the Deschutes Skyline Project. Proposal was revised with letter of support and the landscape was expanded to 257,850 acres (2013).
4 Included funding for FY 2020 in proposal – FY 2010 to FY 2019 total from proposal is $11,844,237. The proposal was revised with a letter of support – the revised total (not including FY 2020) is $10,823,285 (2014).
5 Note that 2012/2013 projects included funding beyond FY 2019 in total funding requested – FY 2012/2013 to FY 19 requests are less.

Size of landscapes, proposed treated acres, and NEPA approved acres

Landscapes varied by size ranging from the original proposal of 130,000 acres on the Deschutes Collaborative Forest Project to the expansive 2.4 million acres of the Four Forest Restoration Initiative. As implementation began, sometimes boundaries were changed to incorporate a larger land area. However, most remained close to their original proposed boundaries. The twenty-three landscape project proposals covered more than 15.5 million acres with an average of just over 700,000 acres. Project goals for treated acres tended to run below, and sometimes substantially below, landscape area covered. For example, the Grandfather project in North Carolina aimed to treat 41,685 acres, about 12.5 per cent of its 330,360 acre landscape. In some cases, this discrepancy was due to the complexity and cost of treatments; in others, landscape boundaries incorporated lands that were not in National Forests and therefore could not be treated using CFLRP dollars. Proposed treatment areas ranged from about 32,000 acres on the Colorado Front Range to more than 600,000 acres in the 4FRI project. Seven of the ten 2010 projects sought to treat over 100,000 acres of National Forest lands. Eight of the thirteen of the 2012 and 2013 funded projects had targets over 100,000 acres, with some over 300,000, such as the Shortleaf Bluestem project in Arkansas and Oklahoma, and the Longleaf Pine project in Mississippi. The smallest three projects from the second group of landscapes aimed to treat approximately 40,000 acres each. Overall, the projects sought to treat around 3.5 million acres on national forest lands within landscape boundaries for an average of just over 180,000 acres per project. This represents about 20 per cent of the land area within the project boundaries of all CFLRP sites.

To treat these acres, the US Forest Service had to first develop, or have completed, Environmental Assessments or Environmental Impact Statements under the provisions of the National Environmental Policy Act of 1969 (NEPA). These required planning documents that can take years to develop, and which may face challenges from stakeholders. Planning work had to be completed for some portion of each landscape before submitting a proposal so that the agency could implement some projects shortly after receiving funding. In terms of area covered by NEPA-required planning documents, the number of acres listed in proposals tended to be significantly smaller than the proposed treated acres for most of the landscapes. One of the key challenges was that CFLRP funding could not be used for planning; thus, forests had to allot other funding to conduct the necessary planning before they could use the CFLRP funds for implementation. For the twenty-two projects, for which we have clear information provided in project proposals, the average project had about 73,000 acres of NEPA documentation completed while the proposed treatment area averaged nearly 180,000 acres. Most of the landscape projects had less than 50 per cent of their treatment area covered by NEPA decisions ahead of their selection for the program. There is wide variability here with a few landscape projects having less than 10 per cent of their proposed treatment area NEPA-ready, and others having more than their proposed treatment area already covered by NEPA documentation and associated decisions.

The discrepancy between NEPA-ready acres and project target acres clarifies why so many of the CFLRP landscape groups spent significant time completing planning documents before engaging in the implementation of projects at a large scale. The 4FRI provides an example where they had approximately 83,000 of their more than 600,000 target acres NEPA-ready. They spent the first five years of the project completing planning documents for a large-scale plan for restoration projects on more than 580,000 acres according to the Record of Decision. The Amador Calaveras Cornerstone project had 1,080 NEPA-ready acres when the project was funded and an additional 18,000 acres in the NEPA process, accounting for nearly half their planned 38,500 treatment acres. The collaborative group was initially uncomfortable with their lack of participation in such a large proportion of the planning for landscape restoration, but ultimately developed processes for review and approval of NEPA documents already completed or in progress. Places where NEPA documentation was complete before CFLRP dollars were awarded were able to move at a quicker pace to achieve their restoration targets. For example, the Colorado Front Range Roundtable project completed their proposed treatments on their target 31,600 acres in around four years into the project. This caused some friction with stakeholders who were not involved in planning those acres for their collaborative project (see Bergemann, Schultz, and Cheng in this volume). In response, the agency initiated environmental impact studies in new project areas to continue to complete restoration during the remaining five years of the program, and involved the collaborative group more thoroughly in vetting this project. The Longleaf Pine site in Mississippi had a unique opportunity with a Categorical Exclusion (CE) from detailed NEPA analysis post-Hurricane Katrina, which led to all 382,000 project acres being NEPA-approved before CFLRP funding was awarded. Thus, implementation could begin landscape-wide in that case.

Ownership and management boundaries

The law that authorized the creation of CFLRP stated that funding had to be spent on US Forest Service lands. Therefore, National Forest lands dominate the program and management responsibility falls on the shoulders of US Forest Service personnel. However, some landscape collaboratives, such as the Tapash, signalled that ecological restoration goals extended well beyond the boundaries of the agency managed lands and covered tribal, non-profit (The Nature Conservancy), and state-managed lands as well. The Kootenai Valley Resource Initiative also took a broader look, as their CFLRP project was only one component of several ecological restoration efforts the collaborative group had developed across land ownership boundaries.

Several of the landscape project footprints cross the boundaries of more than one forest. For example, the 4FRI project in Arizona covers areas that cross the boundaries of four national forests in the landscape, and the Amador Calaveras project in California falls across both the Stanislaus and Eldorado National Forests. Most of the project boundaries only overlap one National Forest; however, nearly all of the projects cover more than one ranger district, even if contained within the boundaries of one National Forest. Projects that cross administrative boundaries require coordination for planning and implementation.

Funds requested

CFLRP was authorized to deliver $400 million over the ten years of the program pending congressional appropriations. With $10 million appropriated in 2010 and $25 million in 2011, the total appropriations are likely to reach closer to $350 million. Each project had to provide evidence that equivalent cost-share funds from the local Forest Service unit or region or partner organizations (funding or in-kind) would support the project as well. The upper target of $40 million for the first ten projects seemed to guide budget decisions for half of the original projects that requested $36 to 38 million for the ten-year timeline. The other half of the original ten project proposals requested between $8–17 million over ten years. The next round of thirteen landscapes only included two that requested more than $30 million. Overall, the projects have requested an average of more than $21 million over the life cycle of the project, for a total of more than $450 million dollars. To make up the difference between funding appropriated under CFLRP and project budget requests, the Forest Service has invested high priority project funding from other sources to support the CFLRP projects. Three of the sites, The Northeast Washington Forest Vision 2020, Ozark Highlands Ecosystem Restoration, and Longleaf Pine Ecosystem Restoration and Hazardous Fuels Reduction projects were all considered high priority restoration areas and were allotted funding from sources outside of the CFLRP allocation, but still operated under the same accountability and reporting metrics of the program.

Collaborative group characteristics

A key aspect of analysis in many of the chapters in this volume relates to the collaborative characteristics of groups seeking to influence CLFRP landscape project planning, implementation, and monitoring. In this section, we briefly summarize characteristics related to the history of collaboration, collaborative partners and roles, agency engagement and roles, and collaborative organization or structure. Specific information for each landscape pertaining to these categories is shown in Table 1.2.

Table 1.2 Summary of collaborative characteristics

Project name	*Collab Group*	*Year formed*	*Interests and Partners Represented*	*Organizational Structures*	*Key Characteristics*
2010 Projects					
Selway-Middle Fork Clearwater Project	Clearwater Basin Collaborative	2008	business, conservation, county govt, state agencies, tribal govt, motorized rec, sportsmen/women, woods products firms	Multilevel Collaborative	operating protocols, specified membership, committee structure, CFLRP work integrated into committees
Southwestern Crown of the Continent	Southwestern Crown of the Continent Collaborative	2009	conservation, smaller collaborative watershed groups, Forest Service, researchers, state agencies, timber industry	Multilevel Collaborative	charter, consensus decisions, committees, specified membership, systematic recordkeeping
Colorado Front Range Landscape Restoration Initiative	Colorado Front Range Roundtable	2004	federal and state agencies, conservation organizations, business interests, recreation users, university researchers	Informal coalition with multilevel characteristics	quarterly meetings, consensus oriented, self-described as "voluntary informal coalition", open membership
Uncompahgre Plateau Collaborative Restoration Project	Western Colorado Landscape Collaborative	2000	federal and state agencies, conservation organizations, business interests, university researchers, electricity providers, other collaborative groups, recreation users, grazing industry and lease holders	Interorganizational executive committee with multilevel characteristics	MOU for the UP Project then WCLC redefined MOU and membership, consensus decision making, open participation, extensive community engagement
Four Forests Restoration Initiative	4FRI Stakeholders Group	2008	federal and state agencies, conservation organizations, county govt, local fire depts, timber/wood products industry, university researchers	Multilevel Collaborative	charter, working groups, decision rules, open membership, systematic recordkeeping

continued

Table 1.2 Continued

Project name	*Collab Group*	*Year formed*	*Interests and Partners Represented*	*Organizational Structures*	*Key Characteristics*
Southwest Jemez Mountains	Southwest Jemez Mountains Collaborative	2009	proposal development: local fire mgmt, tribal govt, state and federal agencies, national lab, economic dev orgs, council of govts, forest industry, university researchers, conservation organizations	Informal coalition with interorganizational executive committee characteristics	small lead partnership committee, participatory planning for proposal but less involvement over time; collapsed and in 2014 restructured to be more formal multilevel collaborative
Dinkey Landscape Restoration Project	Dinkey Collaborative	2009	state and federal agencies, fire safe council, conservation organizations, private landowner, timber mill, researchers	Multilevel Collaborative	steering committee, charter, specified membership, professional facilitation
Deschutes Collaborative Forest Project	Deschutes Collaborative Forest Project	2010; 1993–2004	19 stakeholder groups on steering committee, evenly represented: university researchers, environ groups, local govt, state agencies, federal agencies, recreation users, fire mgmt, timber industry	Multilevel Collaborative	charter, committees, structured membership by interest representation, consensus oriented decision-making process, systematic recordkeeping
Tapash Sustainable Forest Collaborative	Tapash Sustainable Forest Collaborative	2006	state and federal agencies, tribal government, and TNC signatories	Interorganizational executive committee	MOU, executive committee of signatory landowners, less opportunity for broad stakeholder engagement until project-based groups established several years into CFLRP

Accelerating Longleaf Pine Restoration in NE FL	None specified	N/A	state and federal agencies, conservation organizations, private research organizations	Informal coalition	periodic and intermittent partner communications including one-on-one and open communications, stakeholder meeting every 1-3 years for updates
2012/13 projects					
Kootenai Valley Resource Initiative	Kootenai Valley Resource Initiative	2001	tribal govt, county and city govt, conservation organizations, state wildlife agency, local businesses, forest industry, research organizations	Multilevel collaborative	Joint powers agreement between Kootenai tribe, and county and city government, monthly board meetings, many subcommittees, 11 members and numerous partner organizations, systematic recordkeeping
Weiser-Little Salmon Headwaters Project	Payette Forest Coalition	2009	local govt, conservation organizations, forest industry, state and federal agencies, recreation users, ranchers	Multilevel collaborative	Professional facilitation, steering committee, consensus decision making rules, sub-groups, systematic recordkeeping and online forum
Zuni Mountain Collaborative Landscape Restoration Project	Zuni Mountains Collaborative	2005	state and federal agencies, conservation organizations, forest products industry, recreation users, university researchers, tribal govt	Informal coalition	Collaboration centered on information sharing and multiparty monitoring. Includes a growing network of stakeholders. Periodic meetings, workshop, and field trips.
Amador Calaveras Consensus Group Cornerstone Project	Amador Calaveras Consensus Group	2009	federal and state agencies, conservation organizations, county govt	Multilevel collaborative	Community based collaborative, MOA with 26 partner members as of 2011, open participation, monthly meetings, work groups, formal consensus process

continued

Table 1.2 Continued

Project name	*Collab Group*	*Year formed*	*Interests and Partners Represented*	*Organizational Structures*	*Key Characteristics*
Burney Hat Creek Basins	Burney Hat Creek Community Forest and Watershed Group	2010	conservation organizations, local govt, tribal government, forest products industry, contractors, state agencies, fire agencies	Multilevel collaborative	charter, diverse and specified membership, meetings every other month, work groups and technical advisory groups, consensus oriented
Southern Blues Restoration Coalition	Blue Mountain Forest Partners/ Harney County Restoration Collaborative	2005/ 2008	private wood products industry, conservation organizations, local government, ranchers	Multilevel collaborative	two collaborative groups with separate governance structures outlined in manuals, each has professional staff and operational funds, coalition meets every other month, operates in a zone of agreement, has subcommittees
Lakeview Collaborative Landscape Restoration Project	Lakeview Stewardship Group	1998	conservation organizations, private wood products industry, local govt, high school, chamber of commerce, state agencies	Informal coalition	quarterly meetings, consensus oriented, long-standing collaborative relationships, non-profit fiscal mgmt, coordination and planning, strong community engagement, tribal consultation
NE Washington Forest Vision 2020	NE Washington Forestry Coalition	2002	Private timber industry and conservation organizations form the coalition, state agencies, tribal govt, county commissioners and recreation users partner	Interorganizational executive committee	MOU and leadership sub-group, monthly board meetings, consensus decision making

Grandfather Restoration Project	None specified	2010	conservation organizations, state and federal agencies, university researchers	Informal coalition	network of partnerships with quarterly meetings established during proposal development, active partner engagement in project implementation
Longleaf Pine Ecosystem and Hazardous Fuels Reduction	De Soto Sustainable Forest Collaborative	2005	federal agencies, conservation organizations, university researchers	Informal coalition	loose network of partnerships, collaborative team review and field trips
Ozark Highlands Ecosystem Restoration	Oak Ecosystem Restoration Team	2000	state and federal agencies, conservation organizations, private industry, university researchers	Informal coalition	twice yearly meetings, part of the South-Central Fire Learning Network
Shortleaf Bluestem Community Project	Ouachita Mountains Shortleaf Bluestem Alliance	1991	state and federal agencies, conservation organizations, forest products industry, university researchers, private research firms	Informal coalition	part of the South-Central Fire Learning Network, regular meetings, frequent field trips
Missouri Pine Oak Woodlands Restoration	None specified	2009	state and federal agencies, conservation organizations	Informal coalition	periodic meetings, Forest Service primary decision makers, multiparty monitoring projects, technical assistance and outreach work by partners

Collaborative history

In order to be accepted, all landscape project proposals had to demonstrate that the agency and collaborative partners had worked together successfully in the past. Thus, there was a history of collaboration for each of the projects, which was outlined in project proposals. The longest standing collaborative group was the Shortleaf Bluestem project in Arkansas which began collaborative engagement in 1991. Other long-standing groups included the Lakeview Stewardship group on the Malheur National Forest, which was formed in 1998, and the Uncompahgre collaborative in Colorado which was formally established in 2000 but which built on other collaborative activities in the region including the Public Lands Partnership of 1993. This story is not unusual as all projects had to demonstrate prior successful efforts in working collaboratively with stakeholders to accomplish forest management goals. Twelve of the collaborative groups subsumed CFLRP landscape projects into their existing program of works, which were formed well in advance of the program. Eleven of the groups were established between 2008 and 2010 in anticipation of the CFLRP, or, in a few cases, shortly after funding was awarded. These groups aimed to develop a program of work or broaden existing collaborative efforts to encompass the functions and boundaries that would be relevant to the CFLRP more specifically. In every case, collaboration had been happening in the region well in advance of the CFLRP; the program simply created the impetus to adjust or create new structures and processes to incorporate new activities or establish new boundaries for forest management work.

Collaborative partners and roles

There is a great diversity of collaborative partnerships across the CFLRP projects, which reflects the diversity of interests invested in national forest lands and the potentially wide-reaching impacts of forest restoration. Stakeholders come from all sectors. Non-governmental conservation organizations representing environmental, wildlife, and recreation interests, including both national level organizations as well as locally or regionally based conservation groups, are involved in every collaborative group. Multiple levels of government agencies are involved as well, especially state and federal land and natural resource management agencies, but also some county and town officials in many collaborative groups. Academic institutions, especially those with forestry or natural resources programs, play a role in many of the collaborative groups frequently providing science and monitoring support. Private science-delivery firms and consultants are also involved in a few of the collaborative groups for similar purposes. The forest products industry has representatives on most of the collaborative groups, particularly where the timber industry is still active. Native American tribes also participate where landscape boundaries either overlap with or are nearby tribal lands. Each of these groups play a role in representing the values and interests of their constituencies, and supporting collaborative work in a variety of ways.

Non-governmental conservation organizations fill a variety of key roles in collaborative groups. These organizations are sometimes placed-based, non-profit environmental, watershed, or community-based forestry groups. Frequently, representatives from larger national or regional non-profit organizations such as The Nature Conservancy, The Wilderness Society, Wild Earth Guardians, Center for Biological Diversity, and others are involved. Representatives from these organizations are often the most ardent voices focused on environmental interests. Beyond representing key interests, staff from some of these conservation organizations commonly provide support through coordination and administration, development of science and technical information, and building bridges among stakeholders. These participants often convene a group and provide significant support in keeping collaborative processes moving. While conservation organizations play a significant role in the operation of collaborative groups, they also play a role in communication to the public, providing opportunities for information sharing through a variety of tools such as websites and email list-servs, webinars, in-person meetings and conferences, and media outreach.

To provide a few examples of the roles of non-governmental conservation organizations, The Wilderness Society played a convening and coordinating role on the Southwestern Crown of the Continent Collaborative group as it formed. Representatives also helped develop the CFLRP proposal and served as co-chairs for several years. Other conservation organizations, such as the watershed-based groups of the Blackfoot Challenge, Swan Ecosystem Center and the Clearwater Resource Council, all sent representatives who became active participants and held leadership positions on committees and task groups on the Southwest Crown. Other organizations, such as The Nature Conservancy and the Trust for Public Land representatives were also involved early on.

On the Deschutes Collaborative Forest Project, leaders from The Nature Conservancy's Pacific Northwest Fire Learning Network and the Central Oregon Partnerships for Wildfire Risk Reduction played key roles envisioning and creating the collaborative group that brought together five other existing collaborative groups in the region to form a 'super collaborative' to oversee the Deschutes CFLRP project. They were the primary authors of the CFLRP proposal and served as coordinators of the collaborative group for the first several years of the program. Over time, more than seventy organizations (including numerous other non-governmental conversation organizations) have signed on as interested parties or participants in the collaborative process. These examples are not unique; conservation organizations frequently have staff or representatives who play roles of convening, coordinating, or facilitating collaborative groups in the CFLRP. They often provide staffing and logistical coordination behind the scenes to further support the operations of collaborative groups. These organizations provide essential support to collaborative groups while simultaneously representing the values and interests of their organizations.

Science and monitoring needs are demanding tasks that are often filled by non-profits, private contractors, or academic partners. The Southwest Ecological Restoration Institutes in Colorado and Arizona contribute significantly to the CFLRP projects in their respective states. They provide facilitation, coordination, and logistical support roles, as well as scientific and technical expertise to help develop and implement monitoring and adaptive management plans (see Cheng *et al.* and Colavito in this volume). The Southwestern Crown collaborative hired a monitoring coordinator affiliated to the University of Montana to help guide monitoring plan development and project implementation, and worked with a private firm, the Ecosystem Management Research Institute, to provide science and monitoring support. The Accelerating Longleaf project in Florida hired the Tall Timbers Research Station to develop and implement a monitoring program. On the Deschutes project, The Nature Conservancy lent staff support to the monitoring program and helped design a research program to address information gaps and educate partners on research processes and outcomes.

State, tribal, and local government affiliated participants were also actively involved in many of the CFLRP projects. One distinguishing characteristic of these stakeholders is that they are often landholders and land managers of adjacent or nearby conservation lands that are on the landscape, or will be impacted by restoration efforts on Forest Service lands. As a result, the roles of these actors are first and foremost as liaisons and representatives of the interests of their constituencies to ensure that the restoration projects cohere with land management goals in their respective management areas. In some cases, local government officials have stepped into leadership roles, such as in the Deschutes project where a local county commissioner played the role of chair of the steering committee, and in the Kootenai Valley project where a local mayor, a county commissioner, and tribal leader served as chairs on the collaborative board. In most groups, the state, tribal, and local government officials are primarily focused on communication and coordination to ensure that information passes both ways and that local, state, and tribal interests are considered in restoration projects on Forest Service lands. Opportunities sometimes arise to share resources and personnel across agencies to help implement projects.

Agency engagement and roles

As the CFLRP is funded and implemented through the US Forest Service, the relationship of the agency to collaborative groups can be complex. On one hand, the CFLRP opens federal land management to collaborative influence in unprecedented ways by requiring that the agency work with collaborative groups through planning, implementation, and multiparty monitoring, per the legal requirements. However, the central planning and implementation role of the Forest Service creates an imbalance of power between agency staff and stakeholders, and, depending on the history and context of each site, this can create tension within collaborative groups. Moreover, the agency cannot relinquish authority and responsibility to collaborative groups, but must adhere to rules and regulations that limit the ability of agency personnel to engage with collaborative groups under certain circumstances. Still, in most of the collaborative groups, agency personnel have been active participants and contributors to the deliberations of the group.

Agency staff play numerous roles in relation to the CFLRP and the collaborative groups involved in agency projects. Many forests have agency personnel who act as CFLRP coordinators. These staff members serve a coordinating function among Forest Service staff members on specific projects and liaise with affiliated collaborative groups. They communicate with Washington office staff and other CFLRP coordinators around the country to share ideas, troubleshoot, and help manage issues that arise in the program. In some cases, agency staff shared or held a coordinator role on the collaborative group as well. On the Southwestern Crown collaborative, a forest supervisor was one of the co-chairs of the group when the CFLRP project began and, on the Tapash Sustainable Forest Collaborative, a forest supervisor chaired the group for the first several years. Where coordinator roles are less formalized, key agency personnel, often line officers such as district rangers, play a similar function when project activities fall under their purview. For example, rangers on the Amador-Calaveras, Colorado Front Range and the Uncompahgre projects play very active roles in collaborative communication and coordination, serving an important liaison function without intervening in a direct leadership role in the collaborative group.

Butler (2013) assessed the agency role in collaborative groups within the first ten landscape projects during the first two years of the CFLRP and found that agency personnel played four different roles: *leadership* as coordinators or chairs of groups; *membership* serving as voting members on decision-making bodies; *involvement* by engaging consistently and regularly in dialogue and deliberations but not in decision-making; and *intermittent* or inconsistent engagement. This research showed that, given the legal context of agency authority and concerns about the Federal Advisory Committee Act, the agency role in collaborative groups was the least problematic and most productive at the involvement level. In this category, agency staff freely and openly shared information, described procedural and institutional issues, provided data, brainstormed ideas – in short, fully engaged in dialogue and deliberations, but did not participate in the final decision-making or serve a lead role in the collaborative group. Antuma *et al.* (2014) assessed the 2012/13 projects and found that agency involvement in the collaborative groups was moderately high to very high for all of the 2012/13 sites. Agency personnel were involved in a variety of roles, but the role of liaison and communication between collaborative groups and agency leadership was central on most collaborative groups.

Collaborative organizational forms

Collaborative groups working on the CFLRP projects vary in structure and function. In some cases, the collaborative groups incorporated the functions and tasks of the CFLRP projects into their existing structures, either by integrating work into existing committees or by establishing new committees or task groups to get project work completed. Others chose to create new collaborative organizations to focus primarily, if not solely, on the work of their CFLRP projects. Those that undertook this strategy tended to: align the scale of the collaborative to the project boundaries; design a multi-level collaborative structure (i.e. a larger group with steering committees and subgroups or working groups) for deliberation, task completion, and decision-making that focused on project goals; and include a broad array of stakeholders so as to align with legislative requirements. They also drew on prior experience with the collaboration to guide how they structured collaborative activities within the project (Monroe and Butler, 2015).

We grouped collaborative structures into three types: a multi-level collaborative organization; an interorganizational executive committee; and an informal coalition. This section describes each of the three collaborative structures in general terms and provides examples of some of the cases that are featured in analyses in subsequent chapters. We drew on project proposals as well as qualitative case study research to inform our categorizations. One of the challenges in categorizing these groups is that the groups are not static. Thus, some groups changed from one type to another. Furthermore, there are differences within categories that we identify in this section and sometimes the boundaries between one type and another are not entirely clear. In this section, we describe each group sufficiently, so that readers can have a sense of the collaborative organizations of the cases that are studied in subsequent chapters.

Multi-level collaborative organizations

Multi-level collaborative organizations are the most frequent type of collaborative structure among the CFLRP cases. These types of structures include a steering or executive committee and working groups or sub-committees assigned specific tasks. Some examples of this type of group include the Southwestern Crown of the Continent, Deschutes, Clearwater Basin, 4FRI, Southern Blues, and Dinkey collaborative groups. The steering committee either plays the role of agenda setting and logistical support for the larger collaborative (e.g. Dinkey, Southern Blues, and 4FRI cases) or is a large and diverse representative voting body (e.g. Southwestern Crown, Deschutes, and Clearwater Basin Collaborative cases). Most of these groups were established in anticipation of the CFLRP program, or shortly after they received funding.

Two of the groups combined existing collaborative groups to form a new collaborative body. The Deschutes brought together five other collaborative groups focusing on forest restoration and wildland fire management to form what they called a 'super collaborative' shortly after receiving CFLRP funding. They have a decision-making and membership structure that consists of a nineteen-member steering committee of diverse interests and five subcommittees focused on specific tasks and open to all interested stakeholders. The group operates under a charter and aims for consensus, but has a minority reporting mechanism to enable recommendations to move forward to the agency in a timely manner when consensus cannot be reached. Two collaborative groups work with the Malheur National Forest on the Southern Blues project. Based in John Day, Oregon, The Blue Mountain Forest Partners had been working together since the early 2000s and was formalized in 2005. The Harney County Restoration Collaborative is located at the southern end of the project area and was founded in 2008. Both groups have their own decision-making rules and, although they operate differently and independently of each other, they come together under the umbrella of the Southern Blues collaborative project to coordinate their planning and implementation of projects. Also, both have a strong working relationship with the Forest Service.

The other four were established to develop CFLRP proposals or began working on CFLRP proposals shortly after they were established. The Southwestern Crown Collaborative began in July 2009 and was formed from partnerships between conservation, economic and rural development, wood products, and educational interests in the region. These interests were all represented in the collaborative early on, with open and voluntary participation. The group operates under a charter that establishes membership and decision-making rules. The group uses consensus-based decision-making and relies on working groups and subcommittees to get project work done. The Clearwater Basin Collaborative was formed in 2008 at the instigation of Senator Crapo of Idaho and out of discussions between conflicting interest groups within the Basin. By the time of their CFLRP proposal, they had expanded to include twenty-four individuals, representing the interests of citizens, business development, conservation, recreation, woods products, and local and tribal governments. The group functions with guiding operating protocols and has clear liaison and communication functions between agency personnel and collaborative stakeholders. The 4FRI Stakeholder Group began the formation process in 2008 drawing on more than a decade of various stakeholders collaborating in restoration efforts. Representatives from counties, cities, and towns in northern Arizona came together with forest management and restoration programs, research institutions, local tribes, wood products industry representatives, and natural resource management agencies to create a foundation document, a charter, a Memorandum of Understanding between the 4FRI Stakeholder Group and the Forest Service, with guiding documents for implementation. They use working groups for communications, implementation, planning, monitoring, and industry support. The Dinkey Project Planning Forum came together in 2009 and built on successful small-scale collaborative work on the Sierra National Forest to develop a CFLRP proposal. The group expanded to include members representing multiple interests, including government, environmental, fire safety, forestry, industry, landowners, and tribal interests. The group is a consensus-seeking body, with the Sierra National Forest serving as the lead partner and playing a coordinating role. They had the support of professional facilitators from the beginning, who helped the group establish a charter and who planned and coordinated collaborative meetings and guided the decision-making processes.

The voting membership in each of these collaboratives is inclusive of a range of interests identified in the convening phase. Chairs or co-chairs of each working group or committee, and of the steering committee, play a leadership role, but the leadership and decision-making authority tends to be distributed with the chairs playing more of a facilitative than an authoritative role. Each of these groups has developed a charter or operating protocols that clarify their purpose, structure, decision-making processes, and membership rules. In a few cases, they have a Memorandum of Understanding (MOU) to clarify relationships between the collaborative group and the US Forest Service. These collaborative groups all seek consensus as they make decisions, but in some cases, such as the 4FRI, Southwestern Crown, and Deschutes, they specify supermajority voting and deliver minority and majority recommendations to the Forest Service when full consensus cannot be achieved. Overall, these groups maintain a relatively high level of inclusion of a range of interests (although in somewhat different ways) and they are highly formalized in terms of defining their structures and protocols for collaborative engagement.

Interorganizational executive committee

Some collaboratives used an executive committee of core members to serve as the primary collaborative decision-making or oversight body. The Tapash and the Uncompahgre Partnership (later expanded to become the Western Colorado Landscape Collaborative (WCLC)) serve as distinct examples of this approach. The executive committees tend to operate by informal consensus and receive requests and input from stakeholders periodically. The committees were established through a Memorandum of Understanding (MOU) between signatory agencies with land management responsibilities in the region. While the level of participation and input from the stakeholders varies in each collaborative, a limited number of representatives on the executive committee serve a large-scale visioning, information sharing, and, to some extent, oversight role. In each of these cases, the intent was to bring together decision makers of major landholders across the landscape to share information and coordinate joint efforts and restoration projects.

The Uncompahgre Partnership (UP) was formally organized in 2001 with a Memorandum of Understanding (MOU) signed by line officers of the US

Forest Service, Bureau of Land Management, Colorado Parks and Wildlife, Western Area Power Administration, Tri-State Generation and Transmission Association, and Uncompahgre/Com Inc. – a non-profit created to provide administrative services to the partners. After CFLRP funding was awarded to the Uncompahgre project, questions about financial management arose which led to the creation of the Western Colorado Landscape Collaborative (WCLC) and a renewed MOU in 2011, which expanded boundaries and involved several working groups or partnerships. The Uncompahgre Partnership had a long history of broad public engagement which continued under the CFLRP project. The stakeholders involved in the CFLRP project on the Uncompahgre Plateau meet with the Forest Service several times a year on proposed projects, monitoring needs, NEPA scoping, and field reviews to strategize adaptive management. The meetings are facilitated by the Colorado Forest Restoration Institute and use an informal consensus decision-making process. Thus, while the WCLC served as the nominal coordinating body, much of the collaborative work on the Uncompahgre continues to be distributed among a range of organizations and other collaborative groups and projects.

The Tapash Sustainable Forest Collaborative was formed in 2006 with a MOU between land owners and managers across 1.6 million acres, including the Washington Department of Fish and Wildlife, the Washington Department of Natural Resources, the Okanogan-Wenatchee National Forest, The Nature Conservancy, and the Yakama Nation. The collaborative was undergoing a transition in leadership when it was selected as a CFLRP project. In the transition, the executive committee became a place where high-level leaders from the respective organizations gathered to share information, undertake strategic planning, and coordinate restoration efforts. The Tapash approach to bringing in a broader suite of stakeholders to provide input into project design and implementation was much more uneven, leaving numerous stakeholders and mid-level managers within signatory organizations dissatisfied with the level of input they had. The Tapash group sought ways to bring in stakeholders more effectively, trying out new structures and combinations of committees and communication plans several times over the first five years of the program. Over time, the membership of the executive committee changed and the group experimented successfully with having more inclusive project level collaborative groups to serve as the locus of project-based planning and implementation activity, while the executive committee continued to serve as an information sharing and broader strategic planning forum for the signatory agencies and organizations.

Informal coalition

Finally, several of the collaborative groups originally operated as informal coalitions. Informal coalitions, while loosely defined, vary significantly in how they accomplish collaborative work. Some examples that provide a sense of this variability include the Colorado Front Range Roundtable, Accelerating Longleaf Pine Restoration, the Missouri Pine-Oak Woodlands project, the Zuni Mountains project, and the Southwest Jemez Project. The Front Range Roundtable and the Southwest Jemez groups offer somewhat unique examples in this category. The Front Range Roundtable was established in 2004 as 'an informal voluntary coalition', open to all and using consensus seeking processes to come up with restoration principles across a 1.5 million acre landscape (see Cheng *et al.* and Bergemann *et al.* in this volume for more on this case). The group meets quarterly, uses a professional facilitator, and shares ideas and resources across jurisdictions, agencies, and organizations. While they are informal in the sense that their membership is voluntary and open, and there are no firm commitments that arise from the group, they are structured somewhat similarly to the multi-level collaborative organizations. Since being awarded CFLRP funding, the group has established more formal task groups (the Landscape Restoration Team) to conduct monitoring and is now more involved in Forest Service planning activities. The Southwest Jemez organized a Lead Partnership Committee around four core organizations (including agency staff, a conservation organization, and an ecological restoration institute) with a vested interest in restoration of fire-adapted ecosystems in the region. These parties began meeting informally to discuss restoration activities across the landscape in anticipation of the CFLRP program. They undertook an open and participatory process to develop a proposal for CFLRP, but had trouble maintaining involvement with stakeholders afterwards. Thus, while they initially aimed to have an open and informal process, the lead partnership committee became similar to an interorganizational executive committee, until the group reformed under a new charter after the Forest Service withdrew from the partnership committee.

The other three groups are relatively similar in their organization and functions. The Accelerating Longleaf project lacked a well-defined collaborative body. Instead, Forest Service staff communicated with partners one-on-one to get buy-in, garner input, and identify places where the Forest Service and a specific collaborator could work together on aspects of the project. Agency staff maintained open communications with a suite of partners on an ad-hoc basis via phone calls, emails, site visits, and occasional face-to-face meetings with a broad set of stakeholders. The Pine-Oak Woodlands project originally started as a strong partnership between The Nature Conservancy and the Mark Twain National Forest. The collaborative group officially formed in 2009. It is an informal partnership, with meetings on an as-needed basis and no discernible formal structure. Forest Service staff provide most of the leadership and coordination, with key partners providing outreach, communication, and technical assistance. The Zuni Mountains collaborative group was established and is coordinated by the Forest Stewards Guild. Early on, the agency primarily communicated one-on-one with stakeholder groups and funnelled much of their communication through the Forest Stewards Guild lead contact. An annual stakeholder meeting provided an opportunity to the Forest Service and the larger stakeholder group to discuss priorities and projects. Over time, a growing network of stakeholders have become involved, through periodic meetings, workshops, field trips, and multiparty monitoring projects.

Conclusion

This chapter provides a broad overview of the twenty-three collaborative groups, and the restoration projects the agency and collaborative groups proposed under CFLRP. We have outlined some of the key landscape characteristics of each project (Table 1.1) as well as collaborative group characteristics (Table 1.2). Subsequent chapters in the book draw on these cases for their respective analyses.

In terms of landscape characteristics, the CFLRP cases are united by their focus on restoring forests so as to reduce wildland fire threats and management costs, as well as enhance ecological system functions and habitat for target species. However, landscape boundaries vary widely in terms of the footprint of each project and the land ownership and management areas included in each landscape. Projects also differ significantly in terms of the proposed treated acres within each landscape area, planning documents completed, and funds requested for their projects. This variability had implications for the extent to which projects made progress on treatments, the level of coordination they needed to get work done, the amount of planning work that needed to happen to enable restoration project implementation, and the complexity of system conditions that needed to be addressed when defining restoration goals and projects to achieve them.

Collaborative group characteristics were equally if not more variable across the landscapes. While the collaborative groups on each landscape project all shared characteristics, such as the fact that there had been a history of successful collaboration on past projects and that they could bring together a diverse group of stakeholders to undertake landscape restoration, each group was unique in its design, processes, history, membership, and stakeholder roles. The context of collaboration was significantly different in each landscape and that context defined and shaped how collaborative groups were organized and managed as well as the extent to which those groups had an influence on agency decisions and actions pertaining to landscape restoration projects. Many of the chapters in this volume explore this variability to some degree and point to characteristics that clarify how large landscape collaborative forest restoration can be undertaken to increase the pace, scale, and quality of restoration, as well as generate secondary outcomes through collaborative processes.

References

Antuma, J., Esch, B., Hall, B., Munn, E., and Sturges, F. (2014) 'Restoring Forests and Communities: Lessons from the Collaborative Forest Landscape Restoration Program', Master's project, Ann Arbor, MI: University of Michigan

Buma, B. and Wessman, C.A. (2013) 'Forest resilience, climate change, and opportunities for adaptation: A specific case of a general problem', *Forest Ecology and Management*, 306: 216–225

Butler, W.H. (2013) 'Collaboration at Arm's Length?: Navigating Agency Engagement in Landscape-Scale Ecological Restoration Collaboratives', *Journal of Forestry*, 111(6): 395–403

Covington, W.W. and Moore, M.M. (1994) 'Southwestern ponderosa forest structure: changes since Euro-American settlement', *Journal of Forestry*, 92: 39–47

Hessburg, P.F., Agee, J.K., and Franklin, J.K. (2005) 'Dry forests and wildland fires of the inland Northwest USA: contrasting the landscape ecology of pre-settlement and modern eras', *Forest Ecology and Management*, 211: 117–139

Keane, R.E., Ryan, K.C., Veblen, T.T., Allen, C.D., Logan, J., and Hawkes, B. (2002) 'Cascading effects of fire exclusion in the Rocky Mountain ecosystems: a literature review.' General Technical Report. RMRSGTR-91. Fort Collins, CO: US Department of Agriculture, Forest Service, Rocky Mountain Research Station. p. 24

Monroe, A. and Butler, W.H. (2015) 'Responding to a Policy Mandate to Collaborate: Structuring Collaboration in the Collaborative Forest Landscape Restoration Program', *Journal of Environmental Planning and Management*, 59(6): 1054–1072

The Nature Conservancy, Ozarks Ecoregional Assessment Team. (2003) 'Ozarks Ecoregional Conservation Assessment', The Nature Conservancy Midwestern Resource Office, Minneapolis, MN

Westerling, A., Hidalgo, H.G., Cayan, D.R., and Swetnam, T.W. (2006) 'Warming and earlier spring increase western US forest wildfire activity', *Science*, 313: 940–943

Part I

Collaborative governance dynamics

2 Trust ecology and collaborative natural resource management

Marc J. Stern and Kimberly J. Coleman

Introduction

It has become a common dictum in natural resource management literature that trust is critical to grease the wheels of planning, collaboration, and implementation. However, different scholars have conceptualized trust in varying ways. In this chapter, we first review dominant frameworks for conceptualizing multiple aspects of trust and map each framework onto collaborative natural resource management. In doing so, we draw meaningful distinctions between the concepts of trust, trustworthiness, critical trust, distrust, and the actions that trust and distrust can foment. Then, we demonstrate the special utility of the trust ecology framework for collaborative natural resource management contexts and draw upon empirical research from Collaborative Forest Landscape Restoration Program (CFLRP) cases to illustrate lessons and opportunities that emerge from considering its key elements.

Trust, trustworthiness, and key perspectives on each

The most common definition of trust in the literature establishes trust as a psychological state in which one entity accepts vulnerability to another despite the uncertainties and risks associated with doing so (e.g. Rousseau *et al.*, 1998; Stern and Coleman, 2015). This psychological state can be influenced by a range of factors, including: characteristics and disposition of the ***trustor*** (i.e. the individual who is deciding whether or not to trust); characteristics of the ***trustee*** (i.e. the individual about which the trustor is making his or her decision); and characteristics of the context in which the two entities find themselves. The trustor is typically an individual, but the trustee can be an individual, a group, an organization, or even a report or a map.

In recent decades, scholars have given greatest attention to the characteristics of the trustee who enhances the likelihood of being perceived as trustworthy (e.g. Mayer *et al.*, 1995; Schoorman *et al.*, 2007). Perceptions of ***trustworthiness*** are typically driven by the trustor's perception of the ability, benevolence, and integrity of the trustee. ***Ability*** refers to the perceived competence of the trustee within the domain of concern. ***Benevolence*** refers to the trustor's perception of whether the trustee cares about the trustor and has their best interests in mind. ***Integrity*** refers to the trustor's perception of whether the trustee consistently adheres to an acceptable set of principles within the domain of concern. Trustworthiness research provides valuable lessons about the behaviours and communications associated with higher levels of trust or distrust. However, focusing on trustworthiness alone fails to adequately explain other influences on the psychological state of trust, including the role of context and the predispositions of the trustee.

More recently, we have developed the concept of ***trust ecology*** to account for the psychological processes and contextual influences relevant to trust assessments made by people in real world contexts (Stern and Baird, 2015; Stern and Coleman, 2015). The trust ecology framework identifies four forms of trust based on distinct sources: dispositional, rational, affinitive, and systems-based trust. ***Dispositional trust*** is based on individual's pre-existing inclinations toward the world around them. For example, you might be someone who is generally sceptical of anything you hear and of the new people you meet. Your trust must be earned. Alternatively, you might be someone who tends to give people the benefit of the doubt, assuming that most people (or organizations or other entities) are usually trustworthy. In this case, you start from a position of general trust, which can be lost or enhanced through your interactions. Dispositional trust establishes the baseline from which other forms of trust may be built or lost. In natural resource management interactions, high levels of dispositional trust are rarely widespread among the stakeholders who show up to public meetings. More commonly, scepticism and concerns motivate engagement (see Smith *et al.*, 2013). There is not much we can do about other people's levels of dispositional trust or distrust, as these predispositions are developed slowly over a lifetime of personal experience. The other three forms of trust, however, can be influenced through our actions.

Rational trust is based on evaluations of the predicted outcomes of the expected behaviours of others. If I expect you to perform in a way that will benefit me, I am likely to have rational trust in you. These expectations are typically grounded in perceptions of competence based on prior performance, consistency, follow-through, or reputation – similar to the idea of 'ability' above. Building rational trust, thus, typically requires the opportunity to demonstrate expertise, competence, or consistency in actions that align with a trustor's interests.

In contrast to rational trust, ***affinitive trust*** does not involve any sort of specific cost-benefit analysis of expected outcomes. Rather, affinitive trust is based on the development of an affinity for a potential trustee, which can emerge from numerous sources, including the charisma of the trustee, the sharing of positive social experiences, assumptions of similarity in values and backgrounds, or membership in shared social groups or categories, or demonstrations of active listening, caring, or responsiveness. Affinitive trust thus involves positive emotional connections between stakeholders and can be highly predictive of subsequent collaborative behaviour (e.g. Stern, 2008).

Systems-based trust (also known as *procedural trust*) entails trust in the surrounding context or the set of rules that govern interactions, rather than trust in any specific individual or organization. This form of trust lowers the risk of participation if participants believe that everyone will abide by the rules and that the rules are fair. Natural resource governance systems or collaborative forums with certain qualities tend to enhance this type of trust more than others (Ostrom, 1990; Coleman *et al.*, 2018; Fisher *et al.*, 1991). These qualities include:

- Collaboratively developed or agreed upon rules for interaction;
- Procedures that are transparent (easily known) to all participants;
- The development of shared (agreed-upon) criteria for evaluating the quality of potential decisions;
- Agreements about what to do when a participant violates the rules;
- Procedures that protect respectful dissent;
- Procedures for members to collaboratively revise the rules as necessary; and
- Third party facilitation and coordination.

The theory of trust ecology posits that sufficient stocks of all three forms of actionable trust (rational, affinitive, and systems-based) are necessary for long-term collaboration in multi-stakeholder environments (Stern and Baird, 2015). These forms of trust serve as buffers to disturbances to collaboration. Just as diversity within an ecosystem can allow for a quicker recovery after disturbance, diversity of forms of trust can help collaborative entities to recover from shocks. For example, if one entity fails to perform an expected duty, rational trust might suffer. The presence of affinitive trust can enable collaborators to give that entity a second chance or provide support in the form of direct task-related assistance or capacity-building. If a charismatic leader who was the source of high levels of affinitive trust leaves the group, strong systems and associated systems-based trust can ease the transition, as collaborators still know what to do and what is expected of them. As systems or external contexts change, rational and affinitive trust ties can buffer the collaboration while procedures are re-built or revised.

In general, higher degrees of trust within a collaborative network should facilitate better collaboration, as collaborators feel safe to share their ideas, deliberate the ideas of others, and enjoy their interactions with less concern about being mistreated or tricked. However, if trust is too high within a network, there is a danger that complacency can take the place of active deliberation and innovation (Parkins and MacFarlane, 2015; Poortinga and Pidgeon, 2003). If one completely trusts the leaders of the group (or others) to make the right decisions, why bother actively participating beyond the bare minimum? While some disengagement might make planning more efficient, planning outcomes can suffer from a lack of creativity and foresight. If diverse forms of knowledge, ideas, and deliberation fail to surface, adaptive management may stagnate, leading to suboptimal decision-making. This brings us to the equally important, yet distinct, concepts of distrust and critical trust.

Distrust is markedly different from a lack of trust. A lack of trust implies that there is insufficient information for a trustor to formulate a strong opinion about whether to trust or distrust a potential trustee. Distrust, on the other hand, is the expectation that the trustee will act in a way that explicitly goes against the values or well-being of the trustor. Distrust tends to lead to unproductive conflict (e.g. Stern, 2008). In collaborative natural resource management settings, distrust can arise when arguments become personal, when people feel their identities or core beliefs are not respected or valued, when rules and decision-making procedures are unclear or unfair, when people or organizations fail to keep their commitments, or when the costs of management actions feel unfairly distributed.

Critical trust, in contrast, involves a generally trusting stance coupled with a healthy dose of scepticism (Parkins and MacFarlane, 2015; Poortinga and Pidgeon, 2003). Critical trust may develop most naturally from an acknowledgement of the uncertainty inherent in any management action or plan, no matter who is in charge. This recognition can keep people at the table and serve as a catalyst for learning and innovation, rather than a reason for sabotage, subterfuge, or competition. Furthermore, shared and openly acknowledged concerns about uncertainty can help build other forms of trust between participants as they can come to see themselves as working shoulder-to-shoulder to reduce uncertainty and better understand and address concerns, instead of using them in head-to-head arguments. This potential positive function of critical trust depends upon the degree to which other forms of trust have developed. In their absence, scepticism may serve either positive or negative functions. On the one hand, it may sustain participation and extend helpful deliberation. On the other, doubts about the other entities or the process may exacerbate adversarial conditions.

In summary, trust research suggests that individuals who show up to natural resource management initiatives may typically arrive with relatively high levels of scepticism (low levels of dispositional trust). Individuals and organizations who are seen to have high degrees of integrity, competence, and caring tend be perceived as more trustworthy. However, these elements alone do not ensure that trust will develop. If stakeholders feel that others do not hold their best interests in mind or that the rules of engagement are not transparent or fair, trust is unlikely to develop. Our research to date has found that co-developed, agreed-upon procedures for collaboration and decision-making can be a powerful first step, establishing the conditions for trust to emerge (Coleman and Stern, 2018b). Third party facilitators, in particular, can play a powerful role in creating favourable conditions for trust development by creating spaces for stakeholders to not only demonstrate their knowledge and competence, but also to share personal connections through friendly interactions designed to expose each other's shared humanity (Coleman *et al.*, 2018). We share some anecdotes below from our research on

CFLRP cases that demonstrate some of the most important roles of trust we have uncovered in our research within these initiatives.

Lessons from the field

While our perspectives on trust and natural resource management have developed from years of field research in diverse contexts and from extensive literature review, in this chapter we share examples and lessons learned from recent case study research on four collaborative groups from the CFLRP. Over approximately three years, we conducted thirty-three interviews with project participants, made observations during eight site visits (two per collaborative group), and held qualitative reviews of archival documents, including meeting notes, websites, e-mails, and publications (for more detail on methods, see Coleman and Stern, 2018). To protect sensitive information and to comply with the conditions of the Human Subjects Committee of our Institutional Review Board, we do not divulge the specific collaborative groups by name, but rather refer to them as Collaboratives A, B, C, and D. For all four groups, we use the term 'the collaborative' to mean the individuals both internal and external to the USFS who regularly met, deliberated, and planned for implementation of projects within CFLRP landscapes.

Trust ecology

Just as species richness (i.e. the number of different types of species) contributes to an ecosystem's ability to withstand disturbance, so too does trust richness contribute to the resilience of collaborative groups. In other words, each of the three forms of actionable trust can play different roles within collaborative groups, and all three are important for successful long-term functioning. When a collaborative group lacks one or more types, the functioning of the group suffers. We share examples from CFLRP cases we have examined that demonstrate this point.

One collaborative group, Collaborative A, decided to address a lack of systems-based trust by simply narrowing the scope of their work to areas of known agreement. The group enjoyed moderate to high levels of rational and affinitive trust among members at the outset of the formal collaborative initiative. Members had worked together on previous projects and had developed respect for each other's professional competencies, as well as friendships and social connections. This rational and affinitive trust enabled them to work informally together on multiple projects where they clearly shared objectives. Because collaborating on these agreed upon issues was generally easy, the group never formally agreed upon deliberation or decision-making procedures. Rather, they had one immutable arrangement: they agreed not to address areas of high conflict.

All members in the group were aware of sharp disagreement on new road development within the forested landscape. Wilderness advocates stood against the construction of any new roads, while others favoured their construction to accomplish management objectives, particularly in fire-prone areas. Rather than address the issue or develop processes to hash out disagreement, the group opted to ignore the issue altogether, making an explicit decision to only work on projects of known agreement between all parties. One collaborative group member explained, 'What made it into the proposal were ideas that everyone participating agreed on'. Stakeholders referred to this as the '80 per cent rule', referring to the fact that they agreed about 80 per cent of time. Because this kept them plenty busy, they decided not to address the 20 per cent where they felt likely to be bogged down by conflict. This enabled them to carry on in the absence of any established procedures for decision-making, conflict resolution, or membership requirements. Without any procedures in place, no systems-based trust ever formed to create safe spaces for deliberation about contentious issues. As a result, ecologically significant portions of the landscape have been off-limits for any discussion or consideration of their treatment or management.

Participants we interviewed skirted around the issue, suggesting that collaborative meetings would likely devolve into unending and unproductive debates, with the potential to permanently damage relationships if they were ever to tread into the conflict-laden 20 per cent. Their fears came to fruition when a forest plan revision process started on a nearby landscape (outside the landscape boundary of the CFLRP project). Although many of the same players were involved, the process was conflict-ridden and devoid of collaborative problem-solving. Participants noted that some of the ill will generated in that process spilled over into the collaborative group, creating some tension on otherwise agreeable actions. This further cemented the groups' determined avoidance of addressing ecosystem-based management holistically across the landscape.

Collaborative B lacked any formal rules that might have helped to build systems-based trust. Members of the collaborative group did not feel the need to establish such rules because they had been working on the landscape together for many years and felt they had developed sufficient informal norms for how they interacted and worked together. In practice, this meant that members relied on affinitive and rational forms of trust, nested within personal relationships, to move forward with their day-to-day work. Collaborative members, however, commonly reported worries that these trusting relationships could be lost if key individual members got new jobs, retired, or left the group for any number of other reasons. This created apprehension and concerns about long-term collaborative functioning and commitments.

Collaborative C demonstrated how the unequal distribution of trust relationships can also powerfully influence group functioning and outcomes. In this group, affinitive and rational trust were concentrated among the members of a small decision-making body of representatives of six organizations that had signed a Memorandum of Understanding establishing the collaborative group, referred to as the executive committee. This group made decisions regarding the direction and priorities of the larger collaborative group, often with only limited communication from the broader membership. High degrees of interpersonal trust on the executive committee precluded their perception of the need for establishing more formal and transparent procedures for decision-making, larger-group deliberation, or other collaborative processes. Over time, members of the broader collaborative group developed distrust toward the decision-making body, describing it as a 'black box'. One group member commented on how the lack of an inclusive decision-making process had resulted in missed opportunities to build trust and achieve other social objectives, explaining:

> If it's just the [executive committee members] talking to each other, they may be building trust with one another, but it isn't meeting any broader social objective of building capacity within communities, building trust between citizens and governments, and connecting working folks and conservationists and scientists and folks from different disciplines.

At least in part, due to unclear responsibilities and procedures, the collaborative group failed to accomplish concrete tasks in a timely manner. As a result rational trust waned, both within the executive committee and the broader collaborative group. Members of the broader collaborative began to disengage from the group because they began to doubt the likelihood of accomplishing project goals. Moreover, the lack of legitimate procedures for collaboration (systems-based trust) made participants wary of how decisions were being made in the first place. This eventually led to numerous efforts to restructure the collaborative arrangement. However, without addressing these core concerns, the new structures similarly failed to produce the desired results.

In contrast to the cases above, Collaborative D built strong stores of all three forms of actionable trust over time. The group was made up of people who had been interacting with each other around natural resource issues for decades prior to the inception of the formal collaborative under the CFLRP. Affinitive trust ties existed between some members prior to the start of the collaborative, but a history of contentious and litigious relationships had led to rational, affinitive, and procedural distrust among many stakeholders. Once the collaborative was established, the founding members quickly worked to implement a clear and transparent governance structure to address concerns related to procedural distrust. The structure included: a steering committee of six individuals that would set the direction of the group and monitor the process; content area working subgroups to conduct analyses and make recommendations on technical issues to the larger group; a trained consensus-building professional (i.e. a facilitator) who facilitated the process and enforced a mutually agreed upon charter; and a broad collaborative group that would vote on final decisions. The charter also identified specific processes for individuals or groups to air objections or grievances. The facilitator invited all representatives of all stakeholder groups to help write the charter and agree on its content and conditions.

The facilitator encouraged the development of affinitive trust between individuals through a range of activities designed to help people realize shared values and affinities, employing the principles of interest-based negotiation to facilitate difficult conversations (Coleman *et al.*, 2018; Fisher *et al.*, 1991; Fisher and Shapiro, 2005). The facilitator purposefully structured events, such as field trips, to encourage informal information sharing, relationship building, and the development of empathy between stakeholders. In these activities, he focused on helping stakeholders to see merit in the viewpoints of others, enforcing rules about civil discourse and basic kindness, separating people's interests from their positions so that basic concerns could be addressed in place of arguing for specific actions, and leading processes through which common criteria could be developed for evaluating potential courses of action (see Fisher *et al.*, 1991).

The facilitator also supported the development of rational trust by keeping the work on task so that deliverables were produced well and on time. For example, the facilitator coordinated and helped subcommittees explore specific areas of uncertainty and report back to the group on their findings. As individuals demonstrated their competences to one another, rational trust grew and spread throughout the collaborative over time.

The presence of all three forms of actionable trust enabled the group to function at a high level in which members could agree on their shared progress toward meaningful outcomes. What seemed out of reach prior to the collaboration had become a reality – for example, successful forest restoration treatments that were endorsed by the full collaborative group. One stakeholder commented:

> As a whole, what I've seen is that a lot of individuals have come in with … their guard up … but … as time goes on, those people stick around. They open up to a certain degree and look at the collaborative process as a new vehicle for coming up with mutually agreeable solution.

Members felt compelled to promote and defend the collaborative group to their home organizations (Coleman and Stern, 2018a). When the facilitator moved on to another position, the formal charter, and the norms of practice associated with it, as well as the rational and affinitive trust these norms of interaction helped to produce, sustained the group through the transition to a new facilitator, who was able to pick up right where the prior facilitator had left off (see more on this transition below).

Trust and uncertainty: developing critical trust

In cases where distrust is commonplace, people tend to use uncertainty as a weapon. For example, climate-change deniers often point out uncertainty in climate models as arguments against climate action. Most research suggests that climate denial, however, is based on a deep-seated mistrust of the proponents of climate action and perceived threats to the self-definitions of those on the other side of the argument (Kahan *et al.*, 2012). In contrast, when safe spaces are created for conversations or interpersonal forms of trust to develop, the role of uncertainty can change dramatically. Instead of being used as a weapon, uncertainty can become the focus of meaningful conversation about problem solving. With adequate stores of procedural and interpersonal forms of trust, acknowledgement of uncertainty can help to focus counterparts on fruitful discussion about efforts to minimize that uncertainty through new knowledge acquisition, or to explore the means for adaptive management moving forward. In this way, scepticism shifts from doubts about others to concerns about how decisions will impact desired futures. The former leads to unproductive conflict; the latter can lead to productive deliberation.

For example, Collaborative D was tasked with reducing the risk of catastrophic wildland fire on a specific landscape. A recent wildlife survey had revealed that a particular species of concern was present on the landscape. The collaborative group sought guidelines regarding the size of snags in which the animals nested, so as to avoid removing trees that might serve as habitat. However, the group discovered that relatively little was known about this topic. The group worked together to identify scientists who were actively researching the species in question. They then invited those experts to make presentations at collaborative meetings. The group engaged in a series of dialogues to process what they heard from the presenters, as well as the limited additional available science they could find about the species and similar animals. This allowed them to create a matrix that prioritized sites for treatment, as well as harvesting guidelines for those sites – which could then be monitored over time. Several members of the group represented wildlife advocacy groups and they certainly could have used the lack of scientific information to stall or completely derail mechanized harvesting on the landscape if they did not trust the intentions of their co-collaborators. However, interpersonal and system-based trust within the group had unearthed a common interest in promoting a functioning healthy ecosystem. Ultimately, this allowed the group to reach consensus despite scientific uncertainty.

Turnover and transitions

Given the longevity of the collaborative groups involved in the CFLRP and more broadly in natural resources management, turnover and transitions are inevitable. Turnover in key people within a collaborative group can erode trust in multiple ways. Rational trust can be lost when a knowledgeable member moves on and is either not replaced or a new member needs to get up to speed. The loss of a charismatic leader or friend within a group can dampen motivations if the affinitive trust ties to that person heightened enjoyment of the work. In other cases, turnover of personnel can provide a fresh start, particularly if the outgoing individual was poorly regarded. In nearly any turnover situation, systems-based trust can be weakened, as new members may be unaware of the established procedures or group norms that have become taken-for-granted. Actions that go against group norms can lead to rifts in pre-existing patterns of collaboration or associated trust, group identity, and motivation.

In our experience, focusing on the development of clear procedures and associated systems-based trust can help groups to navigate turnover. In Collaborative D, when the facilitator left for another job, the group easily weathered the change because they had a process in place for a smooth transition between outgoing and incoming personnel. Specifically, the group planned for the transition by including an overlapping period between the incoming and outgoing facilitators. During this period, the two facilitators attended meetings together and co-authored correspondences to the broader group. By the time the outgoing facilitator had moved on, the new facilitator fully understood group norms and had built positive relationships and trust with other members through consistent performance and maintaining the positive elements of the work of the prior facilitator described above.

When the coordinator of Collaborative A accepted a new position, the process to hire a replacement was fairly drawn out, leaving the group without a coordinator for close to a year. To make matters worse, the group did not have clear procedures to build systems-based trust other than narrowing the scope to agreed-upon issues. The coordinator had been in charge of scheduling meetings, setting agendas, conducting outreach to update collaborative members, and generally making sure that members were accomplishing assigned tasks on time. Without anyone in that role, the group's work stalled. Meetings became ad hoc and irregular, rather than monthly, and the tenor of the work changed from proactively planning future projects to reactively addressing only emerging issues. As the months went on, members began to doubt that the group would be successful in producing high quality deliverables and outcomes, and rational distrust began to rise. As one member explained, many tasks, including monitoring, came to a halt almost entirely. They elaborated, saying:

> Having someone take the lead on stuff like fieldtrips and activities and that type of thing, it makes them actually happen. If no one's pushing for a fieldtrip or for monitoring ... that's something that tends to fall behind when people have too much to do. Monitoring is kind of the first thing that falls off and so without a coordinator, there wasn't really much monitoring taking place.

Finally, a new coordinator was hired. As work ramped back up toward its previous pace, trust levels slowly rebounded.

In Collaborative C, turnover resulted in an increase in trust and productivity. In this case, it was widely believed by group members that the government official in charge of federal land management was only engaging with other members because collaboration was mandated, and that the engagement amounted to lip service. Members lacked all three forms of trust with this person, doubting that their concerns would actually be heard, considered, and incorporated into forest planning in a meaningful way. When that official rotated to a new position, the collaborative got a fresh start with someone new. As members learned that the new individual subscribed to collaboration wholeheartedly in both a procedural (systems-based) and a personal (affinitive) sense, trust levels began to rise within the group and engagement increased.

In nearly every collaborative group, turnover is inevitable. Without substantial systems-based trust based on mutually agreed upon procedures, interpersonal forms of trust can erode quickly. The establishment of procedures based on core values concerning appropriate engagement and power-sharing can strengthen group identity (Morgeson *et al.*, 2010). When members commit themselves to this identity, the collaborative can better weather member turnover. The real trick to building effective systems-based trust that can persist under conditions of high turnover might be finding the right facilitator, coordinator, or other leader to convince a group that the exercise of establishing a shared charter is worthwhile. Rather than likening the process to negotiating an agreeable contract, in these cases it might be more appropriate to treat it as a cathartic opportunity for members to articulate their core values and strengthen their commitment to the group and its ideals.

Implications for policy and practice

Our explorations into trust have made it apparent to us that trust is a multidimensional concept. We argue that any discussion of trust should address the question, 'What kind(s) of trust?' This question can illuminate potential blind spots and provide strategies for addressing them. We have often witnessed natural resource professionals assuming that people should trust them because they are experts. Such assumptions, which are based entirely on the concept of rational trust, neglect the importance of affinitive and systems-based trust. A more holistic understanding of the different sources of trust (and distrust) can play a critical role in enabling and enhancing the long-term collaboration of diverse stakeholders.

Our research suggests that attention to developing diverse forms of trust broadly within collaborative networks can yield meaningful benefits. It enables risk-taking behaviours and opens up deliberation and information sharing. Trust buffers collaborative groups against internal and external shocks, such as performance failures, turnover, ecological disturbances or catastrophes, or drastic changes in political climate. It makes it easier for the participants to work together on the inherent uncertainties in natural resource management, rather than using those uncertainties as devices of attack. We urge those working in these settings to explicitly focus on creating healthy trust ecosystems in which to work.

In our experience, building healthy stores of diverse forms of trust requires supervisors to value the initiative enough to provide participants with adequate time to get to know each other (affinitive trust). It involves co-creating rules of engagement upon which everyone can agree (systems-based trust). It means following through on promises and performing to the best of our abilities (rational trust). And, it involves facilitation that helps participants learn how to see and openly acknowledge merit in the opinions and arguments of their counterparts, even if they disagree (all forms of trust). Our observations suggest that systems-based trust may form the critical foundation upon which interpersonal trust can flourish (Coleman and Stern, 2018b). This should be encouraging to anyone working in these spaces, as it provides a clear starting point – jointly developing rules and procedures that everyone can agree upon. From there, we have also seen the value of skilled third-party facilitators who can further reduce tensions and move people from face-to-face confrontation toward shoulder-to-shoulder problem solving (Coleman *et al.*, 2018).

We warn groups not to assume they already have enough interpersonal trust to bypass the suggestions we have laid out above concerning the establishment of systems-based trust. Members of both Collaboratives A and B felt constrained by their neglect of establishing clear procedures that could help develop systems-based trust. We recognize that some groups might feel that if everything is going well enough, spending time on developing a charter or specific decision-making rules can feel like a waste of time. Moreover, some might be concerned that the formalization of procedures might cast a shadow on interpersonal trust. Why do we need to codify rules unless we do not trust each other? In short, because interpersonal trust is dependent on individuals, it is subject to loss, via personnel turnover and other trust violations. Without a clear set of agreed-upon procedures, collaborative groups are more vulnerable to multiple forms of disturbance, both from within (e.g. performance failures or disagreements) and from outside the group (e.g. changes in political, social, ecological, economic, or organizational structures or contexts).

The CFLRP included a mandate to collaborate. As such, the external impetus to work together might set these types of collaboratives apart from others that form more organically. The CFRLP cases were externally incentivized from the outset, which likely encouraged some good faith efforts in the early-going to capitalize on the newly available resources offered, possibly in the absence of interpersonal or procedural trust. In other cases, early development of interpersonal trust between opinion-leading participants may have played a stronger role in enabling the first steps toward more collaborative processes. In either situation, developing interpersonal trust relationships early on can initiate broader recruitment, and systems-based trust can enhance the likelihood of continued commitment (Coleman and Stern, 2018b).

Regardless of how collaboratives develop, they typically require significant investment, particularly of time and energy. If future legislation includes similar mandates, it should also include resources to help groups collaborate successfully. Our work has demonstrated the importance of trust for collaborative purposes, as well as a number of strategies to build and support the development of its various forms. At least some of those strategies require significant financial resources (e.g. trained facilitators can be very expensive). As mandates to collaborate grow, policy makers, leaders, and practitioners should consider the resources required for success, particularly the resources required to build trust. In high-conflict areas, this may include substantial financial investments to hire a third-party facilitator to help build trust between members of opposing stakeholder groups. It may also mean significant resources in terms of time (i.e. longer project timelines) to allow for trust building to occur. In short, policy-makers designing programs modelled after the CFLRP should be mindful that trust building is both central to success and likely resource-intensive.

Acknowledgement

This work was funded in part by the USDA Forest Service, Pacific Northwest Research Station, as part of the NEPA for the 21st Century project under agreement number 12-JV-11261976-077, and the Virginia Agricultural Experiment Station and the USDA National Institute of Food and Agriculture McIntire Stennis project 1009027.

References

Coleman, K. and Stern, M.J. (2018a) 'Boundary spanners as trust ambassadors in collaborative natural resource management', *Journal of Environmental Planning and Management*, 61(2): 291–308

Coleman, K. and Stern, M.J. (2018b) 'Exploring the functions of different forms of trust in collaborative natural resource management', *Society and Natural Resources*, 31(1): 21–38

Coleman, K., Stern, M.J., and Widmer, J. (2018) 'Facilitation, coordination, and trust in landscape-level forest restoration', *Journal of Forestry*, 116(1): 41–46

Fisher, R. and Shapiro, D. (2005) *Beyond reason: Using emotions as you negotiate.* Penguin Books, New York

Fisher, R., Ury, W., and Patton, B. (1991) *Getting to yes: Negotiating agreement without giving in.* Second Edition. Penguin Books, New York

Kahan, D.M., Peters, E., Wittlin, E., Slovic, P., Ouellette, L.L., Braman, D., and Mandel, G. (2012) 'The polarizing impact of science literacy and numeracy on perceived climate change risks', *Nature Climate Change*, 2: 732–735

Mayer, R.C., Davis, J.H., and Schoorman, F.D. (1995) 'An integrative model of organizational trust', *Academy of Management Review*, 20(3): 709–734

Morgeson, F.P., DeRue, D.S., and Karam, E.P. (2010) 'Leadership in teams: A functional approach to understanding leadership structures and processes', *Journal of Management*, 31(1): 5–39

Ostrom, E. (1990) *Governing the commons: The evolution of institutions for collective action.* New York: Cambridge University Press

Parkins, J.R. and McFarlane, B.L. (2015) 'Trust and skepticism in dynamic tension: Concepts and empirical refinements on the mountain pine beetle outbreak in Alberta, Canada', *Human Ecology Review*, 21(1): 133–153

Poortinga, W. and Pidgeon, N.F. (2003) 'Exploring the dimensionality of trust in risk regulation', *Risk Analysis*, 23(5): 961–972

Rousseau, D.M., Sitkin, S.B., Burt, R.S., and Camerer, C. (1998) 'Not so different after all: a cross-discipline view of trust', *Academy of Management Review*, 23(3): 393–404

Schoorman, F.D., Mayer, R.C., and Davis, J.H. (2007) 'An integrative model of organizational trust: past, present, and future', *Academy of Management Review*, 32(2): 344–354

Smith, J.W., Leahy, J.E., Anderson, D.H., and Davenport, M.A. (2013) 'Community/agency trust and public involvement in resource planning', *Society and Natural Resources*, 26(4): 452–271

Stern, M.J. (2008) 'The power of trust: toward a theory of local opposition to neighboring protected areas', *Society and Natural Resources*, 21(10): 859–875

Stern, M.J. and Baird, T.D. (2015) 'Trust ecology and the resilience of natural resource management institutions', *Ecology and Society*, 20(2): p. 14, www.ecologyandsociety.org/vol. 20/iss2/art14/.

Stern, M.J. and Coleman, K.J. (2015) 'The multi-dimensionality of trust: applications in collaborative natural resource management', *Society and Natural Resources*, 25(2): 117–132

3 Navigating accountability tensions in collaborative ecological restoration of public lands

Rachel A. Christensen and William H. Butler

Introduction

With the creation of the Collaborative Forest Landscape Restoration Program (CFLRP), the US Forest Service and collaborative groups working together on large landscape restoration projects faced new challenges related to accountability. While collaborative governance has become widely practiced in natural resources management, it has been criticized for its potential to alter systems of public accountability designed to protect the public interest and minimize collusion and corruption (Wondolleck and Yaffee, 2000; Emerson and Nabatchi, 2015). Legal and bureaucratic mechanisms aim to hold public servants to account as they implement laws passed by the legislative branch (Romzek and Dubnick, 1987). Collaboration is often seen as operating outside of this system of oversight and therefore holds risks for public accountability.

And yet, CFLRP mandated collaboration throughout the life of restoration projects, including collaborative participation in creating and implementing a multi-party monitoring and accountability process. A new requirement for collaboration for public lands management increased the magnitude of tensions between bureaucratic and collaborative accountability systems. In this chapter, we draw on data from five CFLRP landscape projects to explore the tensions that arise between bureaucratic and collaborative accountability systems, and describe the ways in which participants in collaborative processes can soften these tensions.

Collaborative environmental management and accountability

Accountability, in simple terms, is how we hold one another answerable for actions. Public accountability is a process or set of mechanisms that ensure answerability by public actors in a democratic society (Bovens, Gooden, and Schillemans, 2014). In their seminal work on public accountability, Romzek and Dubnik (1987) differentiate four types of accountability: bureaucratic, through rules and hierarchical procedures; legal, through laws and regulations;

political, through stakeholder participation; and professional, through standards, norms, and expertise. Within federal agencies, bureaucratic and legal systems are the most influential (Romzek and Dubnik, 1987) where accountability is accomplished through hierarchical reporting, supervisory oversight, and external auditing (Romzek, 2014).

Romzek has also been among the leading scholars examining how public accountability operates differently in public management networks, collaborative groups involving government agencies, non-profit organizations, private corporations, and the public at large (Romzek *et al.*, 2012). In such settings, where lines of authority and responsibility are blurred, collaborative or network accountability becomes a key dynamic within the system and can extend accountability beyond the bureaucratic systems of representative democracy (Hahn, 2011). These systems of accountability between collaborative groups and public agencies have been described as complex, potentially abstruse, and complicated by multiple and competing perspectives on to whom and for what collaborative groups are accountable (Bryson, *et al.*, 2015). The mechanisms of accountability in collaborative systems depend more on informal or relational dynamics among the participants, rather than formal rules and procedures (Romzek *et al.*, 2012; Jedd and Bixler, 2015). When collaborative groups interact with agencies, the system of public accountability is augmented to include what Romzek *et al.* (2012) call informal accountability that operates within interpersonal and interorganizational networks. We adopt the term 'relational accountability' to describe these dynamics, which depends on social interactions, social pressures, and relationships to ensure that group norms and expectations will be met by the participants (Moncrieffe, 2011). Relational accountability forms the foundation and shapes the character of collaborative accountability as social actors interact within the collaborative system and between the collaborative and bureaucratic systems. To be successful in navigating this complex terrain, participants must operate in a context where behaviours and relationships grounded in trust, reciprocity, and respect can be developed and nurtured.

Accountability in public lands management

In public lands management, federal agencies have the primary responsibility and authority for the planning and implementation of projects. Large landscape-scale forest restoration involving public lands is highly influenced by the legal and bureaucratic systems of accountability that dominate public lands management agencies. The US Forest Service was created during the Progressive Era when calls for efficiency, scientific management, bureaucratic organization, and democratic accountability emerged as the key organizing principles of the US Federal Government (Hays, 1999; Behn, 2001). The agency operates under more than 100 laws that both enable and constrain agency actions. Simultaneously, the Forest Service is held accountable to the public through comment and review associated with planning processes required under the National Environmental Policy Act (NEPA) of 1969 and legal review under the Administrative Procedures Act (APA) of 1946. Within this context, agency legitimacy relies on scientifically defensible actions under a hierarchical bureaucratic system which shape its accountability practices.

Over time, agency decisions and actions have been opened to broader scrutiny and oversight with the rise of more collaborative approaches to land and resource management (see the Introduction to this volume). While collaborative environmental management has been shown to be an effective approach to overcoming gridlock, developing innovative ways to solve problems, and defining collective or shared goals (Wondolleck and Yaffee, 2000; Innes and Booher, 2010; Margerum, 2011), it adds collaborative accountability dynamics to agency and organizational accountability systems. Weber (2003) notes that collaborative accountability is shaped by formal structures and processes including facilitated dialogue, decision-making rules, membership roles and responsibilities, standard operating procedures, and enforcement processes among other factors. Simultaneously, there are informal dynamics such as norms, shared perspectives and frameworks, facilitative behaviours, trust, reciprocity, informal monitoring, and rewards and sanctions that drive accountability (Romzek *et al.*, 2014; Weber, 2003).

Rather than changing the underlying system of accountability, authority, and responsibility for managing public lands, collaborative accountability becomes an overlapping system. There are potential tensions within collaborative groups based on differing accountability practices, different organizational goals for participants, and the previous history among participants within the collaborative group. For example, tensions can emerge in collaborative groups due to agency focus on formal oversight and rule compliance that has the potential to undermine trust built through interpersonal relationships (Goldsmith and Eggers, 2004). As Romzek (2014, p. 309) suggests, there is a risk that 'different accountability arrangements can work at cross purposes and undermine the overall process of achieving answerability for performance.' As such, the bureaucratic accountability pressures that agency staff face may be a contributing factor to increasing tensions within collaborative groups working on public land management. Moreover, Romzek *et al.* (2012) find that tensions can emerge between collaborative participants due to competition for recognition or 'turf' staff turnover, and financial pressures for organizational survival. These challenges can be overcome through shared norms of trust and reciprocity, facilitative behaviours of effective communication, recognition of mistakes and actions to fix them, keeping commitments, and sharing information freely (Romzek *et al.*, 2012), as well as relationship building and champion behaviours (Romzek *et al.*, 2014). This ground-breaking work on informal accountability in networked governance lays the foundation for our exploration of how accountability operates within collaborative public lands management, where authority and responsibilities are clear and the calls for collaborative governance are widespread and increasingly mandated (Jedd and Bixler, 2015; Cheng, 2006; Monroe and Butler, 2016).

Methods and cases

This chapter seeks to respond to two key questions: (1) What tensions arise in collaborative public lands management between bureaucratic and collaborative accountability systems; and (2) What are the ways in which collaborative groups and their members navigate these tensions? CFLRP offers a useful context for assessing collaborative accountability, as it: includes cases operating within the same policy context; mandates collaborative engagement throughout planning, implementation, and monitoring; and states that interested parties will participate in 'a multiparty monitoring, evaluation, and accountability process' of the projects (16 USC §7303(g)(4)). The fact that the underlying accountability system remains in place is essential for the democratic legitimacy of this process. And yet, there are potential barriers to broadening the accountability system to include collaborative group input, and CFLRP cases can provide examples for how to navigate these tensions.

We selected five cases from the first ten CFLRP landscape projects, due to their variability in size, location, collaborative structures and processes, and the number of forest management units, and because we had the most extensive datasets on them from our larger research projects. We chose the Southwestern Crown Collaborative (SW Crown), Southwest Jemez Mountains Collaborative (SW Jemez), Dinkey Collaborative, Deschutes Collaborative Forest Project, and Tapash Sustainable Forest Collaborative for our cases (see Butler and Esch for more information on the cases). The SW Crown and Deschutes were organized as multi-level collaborative organizations with a steering committee representing diverse interests and subcommittees focused on specific tasks and open to all participants. The SW Jemez and Tapash collaboratives had more narrowly inclusive groups with a small group of key managers from agencies and conservation organizations who made most of the major decisions about project activities. The Dinkey collaborative was somewhere between these models, with a small executive committee, but multiple task groups and professional facilitation to ensure an inclusive process. These distinctions helped broaden the variability across these cases so that we were able to identify a wide range of barriers and ways to navigate them.

We utilized multiple sources of data, drawing from research project teams from Florida State University and Virginia Tech, and a data sharing agreement between the two. While the two projects varied in terms of their thematic focus, both collected data that pertained explicitly and implicitly to accountability. We analysed eighty-six interview transcripts (fourteen to nineteen per case) and site visit observational records from sixteen landscape or regional level meetings that described collaborative interactions and recorded the content of informal interactions with numerous participants not included in our interview count (see Table 3.1). The project team conducted interviews beginning in the fall of 2011 and continued through the summer of

Table 3.1 Interview and observation record

Landscape Collaborative	*Agency interviewees*	*Non-agency interviewees*	*Landscape and/or Regional Meeting Observations*
SW Crown	5	14	3
Dinkey	7	11	1
Deschutes	3	11	4
Tapash	7	11	5
SW Jemez	9	8	3
Totals	31	55	16

2016 for a longitudinal perspective. We selected interviewees by contacting key participants such as conveners, coordinators, facilitators, and co-authors of project proposals. Then, we expanded our sample by asking these first groups of interviewees to identify other stakeholders and agency staff who were engaging regularly in the projects and collaborative groups. Observations, conducted primarily between 2012 and 2015, afforded the chance to incorporate a wider range of participant perspectives and expand our sample further.

Our analysis used qualitative coding techniques to draw out common and unique themes related to accountability (Charmaz, 2006). The lead author coded all interview transcripts and observation memos using line-by-line coding techniques to identify key tensions and activities that seemed to mitigate some of the tensions (Charmaz, 2006). The second author reviewed those codes and then we independently grouped codes into new categories to build theoretical interpretations of accountability mechanisms, interactions, tensions, and strategies to overcome tensions within and across cases. Then we cross-referenced new categories to find synergies between our coding schema. To protect informant confidentiality, we provide generic affiliations to differentiate between types of interviewees.

Tensions in agency-collaborative accountability

Members of collaborative groups within CFLRP recognized that the ultimate authority for decision-making rested with the Forest Service – as one stakeholder put it bluntly, 'we're not in charge here, the Forest Service is in charge'. Still, the collaborative groups had a level of legitimacy, given that collaboration was required in CFLRP and most groups had emerged over years of working together on the landscapes. Tensions arose when the accountability systems in the agency and the collaborative group did not align well or were undermined by the pressures that dominated the system – namely formal regulatory and bureaucratic accountability. Interviewees identified three key sources of tension repeated in all five of our cases: turnover, bureaucratic procedures, and capacity limitations.

Turnover

Turnover hinders accountability within collaborative groups and between the Forest Service and the collaborative groups as it undermines relational accountability. Because CFLRP projects are slated for ten years of implementation and fifteen years of monitoring, the same people who start on a project will rarely remain throughout. Turnover among Forest Service staff challenges relationships and agreements between the collaborative and the Forest Service, as it limits the capacity of the agency to answer to the collaborative group when new staff come on board. Simultaneously, collaborative accountability is largely relational and informal, and information about collaborative engagement is often held by participants. Thus, turnover amongst collaborative members makes it difficult for the group to maintain momentum and requires effort to rebuild relationships with new participants.

Turnover of Forest Service staff is built in to the agency's bureaucratic system. Turnover partly stems from the longstanding practice of moving personnel between forests to ensure objectivity and to minimize collusion (Kaufman, 1960). Forest Service staff departures have been numerous among CFLRP projects and were identified as an issue in all of our cases. Many collaborative participants complained that the transience of agency personnel inhibited getting things done. Our interviewees pointed to numerous staff positions changing, new forest supervisors cycling through, and a lot of 'actings' (i.e. temporary personnel) filling in. A Dinkey Forest Service staff member noted, 'turnover is not great, just because you have to re-engage and get to know and understand another person'. The key issue is that when someone departs, with whom other participants have a relationship, the participants struggle to recover relational accountability. Often, agency staff fall back on adhering to the spelled-out rules and procedures of the agency, which can undermine relationships with collaborative members.

Non-agency collaborative participants also changed over time. Several collaborative groups experienced changes in collaborative leadership and the group had to recover their footing with new facilitators, new coordinators, even a restructuring to account for the change. In all of our cases, the turnover of stakeholders stalled progress and it took time to reconstruct lines of communication, build trust and relationships among new parties, and re-establish a process for engaging in dialogue and making decisions. A Forest Service staff member concluded,

> we talk about this a lot ..., especially in this context of collaboration that is fundamentally built on relationships. You have these stakeholders who really get to know each other over decades, but changing a decision-maker and landowner can undermine the process.

Bureaucratic procedures

The second major tension was bureaucratic procedures, noted as problematic by respondents across all collaborative groups. A regional Forest Service staff member observed, 'There is that sort of tension set up between internal agency procedures and what the collaborative might want.' The agency has bureaucratic responsibility for managing the land, which comes with related accountability practices. Moreover, there are bureaucratic pressures within the agency for staff to participate in the collaborative process regarding the management of that land. The hierarchical structure, agency culture of professional legitimacy, and overlapping legal requirements led to various tensions between agency and collaborative group interactions and accountability.

Agency hierarchy creates tensions as Forest Service personnel have to answer to supervisors and meet professional expectations, while also trying to engage in collaborative agreements and relationships. Collaborative partners had trouble understanding who within the agency they should speak to and whether that person had the authority to make agreements on behalf of the Forest Service. A Dinkey collaborative member noted that while she felt she could trust individuals, 'as an agency, I don't trust the Forest Service because there's a lack of clarity about who's in charge and how they're going to respond....' Another stakeholder from Tapash said 'I think we can all agree that at times we get frustrated with the Forest Service process. We're working with some great staff, but it is the process itself, that one time or another is going to frustrate us.' A SW Jemez member reflected 'I think these [collaboratives] are made or broken based on Forest Service staff and whether they are willing to stand up to political pressures or pressures from their superiors.' Bureaucracy limits the capacity of agency staff to engage with the collaborative group and uphold agreements, leading stakeholders to be sceptical about agency commitments. Forest Service staff feel pressures that can inhibit their engagement in the collaborative group in productive ways. While they may make agreements, they may not ultimately have the authority to ensure that those agreements are adhered to. They are equally likely to express uncertainty about their ability to make agreements in the first place. Stakeholders become frustrated by the lack of clarity in how the agency makes landscape-level restoration decisions.

Furthermore, Forest Service procedures can be difficult to follow because the agency operates in a relatively insular manner. As one Deschutes member described interactions with the agency, '... the Forest Service is set up to be a technocracy ... experts go in a closed room, close the blinds, run their models, look at the maps, and come up with the answer.' Part of this culture is reinforced by overlapping legal mandates that shape agency engagement in collaboration. For example, the Federal Advisory Committee Act (FACA) establishes rules on how a federal agency can work with collaborative groups through the establishment of an advisory committee (for a full examination of this topic, see Butler, 2013). Because of different interpretations of FACA, agency staff are frequently risk averse and err on the side of not participating fully in collaborative processes. This became a contentious issue in several of the landscapes and led to changing the collaborative organizational structure on the SW Crown and to the eventual collapse and then reorganizing a new collaborative group on the SW Jemez (Butler, 2013).

Limited capacity

The third category of tensions was capacity limitations, especially related to staff time. Forest Service staff have many tasks and their time is divided among many projects, which limits their capacity to participate in collaborative processes. Simultaneously, collaborative partners note that many stakeholders are not paid to participate and volunteering time can be difficult, especially for non-profit organization staff. Moreover, because the CFLRP did not provide funds for collaboration and planning, many groups had to seek other sources of funding for planning and collaborative work, which come with their own accountability responsibilities.

Both agency staff and collaborators identified a lack of available time as a major tension. SW Jemez members, both agency personnel and collaborators, noted that agency staff have 'too many jobs' and a stakeholder specified that 'the problem is that the Forest Service has multiple projects going. What that meant was that [a key staff member's] time was divided.' A partner on the SW Crown further explained:

> There are two capacity issues. One is within the Forest Service. Their capacity issue has to do with the staff, and they really don't have enough time. Then there's the partner capacity issue.... It takes a lot of time and energy to stay at the table with direct expense to participate and you're taking time away from other things.

Thus, without extra funding to provide support for staff members to participate in collaborative activities, it became increasingly difficult to engage in the collaborative process. Additionally, especially for agency staff, bureaucratic accountability pressures ensured that other aspects of their work received attention when time constraints arose, further limiting their participation in the collaborative process.

The second component of capacity was how funding was allocated under the program. One collaborative member summarized an oft-repeated sentiment, saying,

> When everybody signed up for this thing, the idea was that it was supposed to be new money, that is how the legislation is written. That is not how it worked.... It turned into a rearrangement of their existing resources.

The allocated funds could only be used on implementation of projects. Funding for planning, collaborative meetings and tasks, field excursions, and other collaborative activities had to come from other sources. Numerous stakeholders from all collaborative groups affirmed this challenge, summed up by a SW Crown stakeholder who said, 'There's been some frustration that [funding] can only be used for implementation and monitoring and that it can't be used for planning because you can't implement and monitor if you haven't planned.'

These limitations have three key implications for accountability within CFLRP groups. First, accountability from the Forest Service to the collaborative group may have been secondary to tasks that had higher bureaucratic pressures. Second, time constraints for all collaborative participants meant that 'within-group accountability' (which is largely relational) suffered when time was limited and the participants had powerful accountability pressures from outside of the collaborative. Third, in order to implement projects the agency had to complete planning processes and documents. Meeting those accountability requirements without funding from CFLRP meant that both agency staff and collaborative partners had to spend some of their available time seeking other funding sources to fill the gap.

Enhancing agency-collaborative accountability

The most significant accountability tensions in CFLRP arise in relation to how the bureaucratic system influences agency processes and practices in ways that can inhibit engagement with collaborative partners. We found that the following activities have the potential to soften tensions while simultaneously strengthening agency-collaborative accountability: building relationships, listening for understanding, including all voices in decision-making, engaging in effective communication, and fostering group education. These activities build stronger relational accountability, which allows for greater attention to collaborative accountability and offsets some of the pressures of the bureaucratic accountability system that dominates in the agency.

Building relationships

Relationships are the foundation of collaborative accountability. Two activities that support relationship-building are sustained regular interactions and the seeking of common ground. Sustained and repeated interactions can be enabled by an effective collaborative process design that creates regular face-to-face meetings and assigns tasks to subgroups between meetings. Seeking common ground enhances relationships by demonstrating what stakeholders and agency staff hold in common, which can then foster affinitive connections. As relationships strengthen, trust can increase and serve as a glue for collaborative engagement.

Repeated interactions over an extended time period serve as the foundation upon which relationships can be built. According to an agency staff member from the Dinkey project:

> We've spent a lot of time together and of course we are working to promote our shared objectives in sometimes difficult and confrontational situations.... Our relationships have definitely gotten stronger. I think we have become more effective working together not only in this setting but in other settings.

On the SW Crown, several collaborators noted that the collaborative group focused on members spending time together so that they interact with honesty, good will, and good intentions. An agency staff member described the sentiment eloquently, saying:

> It's gotten to a point with the collaborative – where I felt like these groups have our backs no matter what. We're like a family. Working for this agency is hard. There are so many days where I'm just pounding my head against the wall. But you go to these meetings, and because there is such a commitment and desire to succeed, they literally move boulders out of the way to make things happen. There is such a sense of family that when we get together it recharges my battery and reminds you of why you do the job you do. It gets you re-energized to the point of being ready to go to battle again. It never feels like our decision is separate from their recommendation. It just feels like we're just so committed to this end result that it feels like we're working hand in hand.

The importance of these relationships cannot be overstated. They facilitate working together more effectively, but they also ensure that collaborative group recommendations have a greater potential of being adopted and implemented by the agency in the process, which is the essence of collaborative accountability between the group and the agency.

Collaborative accountability can be further enhanced as groups build iterative agreements and then draw on those agreements as social capital for more challenging decisions. By working on the issues where they have identified common ground and zones of agreement, collaborative members and agency staff see commonalities and develop greater affinity ties over time. On the SW Jemez, a Forest Service staff member observed that regular interactions among the core members 'definitely stimulated ideas that we could use by hearing what everybody else was doing. And then you could also see all the commonality.' They then understood where actionable steps could be easily taken within the zones of agreement. Dinkey project group members focused on less conflict-laden topics to build collective capacity to engage in dialogue and reach agreements. After working through a joint fact-finding exercise, one collaborative partner observed that stakeholders had built a strong foundation for future agreements by

> making a choice in that first year to stay out of the high probability [endangered species] areas and to focus on some pieces where there was less conflict because people wanted to get something done and wanted to keep going without spoiling things.

That exercise went a long way to strengthen relationships and thereby collaborative accountability. Another collaborative participant observed, 'even though Dinkey is a very challenging place because of the complicated nature of the issues, it's that beginning experience that keeps us talking to each other.' On the Deschutes, they approached seeking common ground systematically in their deliberative processes. One collaborative partner observed, 'We try not to paint it in terms of negotiation but more in terms of we have different values, and the ultimate recommendations that we come up with are trying to find common ground among those values….' This process enabled the group to make progress on identifying that common ground and strengthen the quality of relationships within the group.

Building strong relationships can help build trust and soften accountability tensions through deepened ties and increased social capital, which serves as a bridge when participants are coming from different perspectives. One of the district rangers on the SW Crown noted that relationships can serve as a cohesive tool, as 'the glue that sticks everything together'. A Dinkey project agency staff member similarly noted:

> I think trust develops as you work with people and you get to know them and you understand their point of view. You get to the place where maybe you don't agree with it but you can accept that they'll always develop compromises that work. So, when you have that trust and when people say things you tend to listen to them, you tend to believe them and that just makes the collaboration work better.

By spending extended time together, working on challenging issues, and clarifying their commonalities, the participants in these collaborative groups not only developed stronger relationships, but also fostered trust which helped them hear each other, believe each other, and, therefore, stay answerable to each other on a different level. In short, building stronger relationships through extended and ongoing interactions and seeking common ground facilitates relational accountability within the group and between the group and participating agency personnel.

Listening for understanding

Related to relationship building, collaborators point to listening for understanding as fundamental. Effective listening helps to clarify accountability pressures that participants are under, creating the potential for empathy. A Deschutes Forest Service staff member shared, 'It's really a discovery exercise for a lot of these folks to understand the complexities that our employees have to respond to and analyze.' Another clarified that this understanding translates into fewer surprises, explaining:

> This collaborative really understands and recognizes that the federal land managers still have decision-making responsibility for the public lands and they're providing recommendations for them to consider. By agency folks being involved in the subcommittees, we can provide some agency perspective. If there's questions about what we can or can't do legally, we're able to advise and provide a little context. From the agency perspective, I think we have pretty good certainty that we'll get recommendations from the steering committee that are things that fit in with our management responsibilities. I think that dialogue is important.

Such dialogue provides space for agency staff to acknowledge their constraints and resources, and collaborative participants are able to understand those limitations in a new light. Once daylighted, they have greater understanding about constraints and mutually supportive actions as they work together on restoration.

Listening creates opportunities to come up with creative ways to approach restoration and broadens perspectives that can enhance collaborative accountability. One Tapash stakeholder reflects that listening allows them to

> come up with creative ideas, asking 'What if, what if' and [the new ideas] can be incorporated somehow into the project. It may turn a project that may be appealed, maybe even litigated, into a project that we can all live with.

A Deschutes community leader noted that 'We have a core group that is willing to be open and listen and work together.... So, we have that understanding and learning that we never had in the past and that has built understanding and education and trust.' These opportunities for listening create space for collaborative members to break out of preconceived ideas and recognize the accountability pressures faced by others, and the implications of those pressures. Through listening, participants were able to create new possibilities together and strengthen collaborative accountability while pre-emptively addressing issues that otherwise would have to be sorted out though legal systems of accountability. Listening was also linked to building understanding and trust – in other words, listening helped strengthen relational accountability among both agency and non-agency collaborative participants.

Including all voices in decision-making

In terms of decision-making, all of the collaborative groups sought consensus before forwarding recommendations to the agency. This consensus-oriented approach welcomed all voices into the process, which can make decision-making cumbersome but leads to more supported outcomes in the final decision. When consensus could not be reached, several collaborative groups found that using minority reports enabled them to provide recommendations to the agency in a timely manner while honouring dissenting opinions about restoration recommendations. By keeping a record of dissenting opinions and including them in recommendations, the group retained accountability to all its members, allowing all voices to be heard even when consensus was not achieved.

The Deschutes used this strategy for several sets of restoration recommendations regarding specific types of ecosystems. Whenever a dissenting opinion emerged, dissenters were asked to write an alternative recommendation that might be acceptable to other members. Then, the steering committee considered alternatives and deliberation continued. If consensus was unattainable, they developed a minority report. One collaborative member described the importance of these reports which were 'forwarded along with the majority opinion to the steering committee and the vote is taken there. Both minority and majority opinions are sent to the Forest Service so they see all sides of the collaborative'. On the Dinkey project, final reports provided clear recommendations while articulating a range of concerns according to interviewees from both the agency and non-agency affiliated members. Two Forest Service staff members noted that 'We're working with a consensus process that allows a certain amount of reservations on the part of its members' and 'the minority opinion will go forward to the Forest Service as will the majority opinion.' Similarly, on the SW Crown, they used a decision-making process that included working to resolve differences through ongoing deliberation, but when there were distinct opinions, a Forest Service staff member noted that:

> They've decided to document the different opinions and present that to the Forest Service for their consideration. There isn't the opportunity for any one individual or individuals to sideline a recommendation. They'll move forward with their interests well-described and explain why they couldn't live with that recommendation to the Forest Service.

Holding space for minority and majority opinions to be communicated to the agency allows stakeholder legitimacy to be maintained and collaborative accountability strengthened, but not at the expense of holding back recommendations and meeting the accountability demands and timelines of the agency.

Engaging in effective communication

Communication processes are central to building relational accountability systems that ensure effective collaborative interaction. Communication makes explicit what might otherwise be implicit or invisible. In short, it ensures transparency. By making discussions and decisions explicit, stakeholders were clear what the collaborative was considering and deciding, thus increasing accountability within the collaborative, between the collaborative and the agency, and to external stakeholders.

Internal communication between the agency and collaborative was handled in various ways by each group. Dinkey and SW Jemez stakeholders described the importance of 'framing' meetings and conversations, telling the story of the collaborative and the work they had done to remind participants of the agreements they made, their accomplishments, and to not rehash resolved issues. The Deschutes and SW Crown both integrated agency personnel on all committees so they were active participants in the discussion. A collaborative participant on the Deschutes described a key function on this internal communication:

> The … committee is going to make sure the collaborators don't feel like there is a disconnect between what they thought they told the Forest Service and how it actually gets rolled out on the ground … to bridge the two sides from planning to implementation such that we make sure those people … felt like the Forest Service heard them … to make sure that sense of satisfaction and trust doesn't get lost.

Strong internal communication between the agency and collaborative group ensured that recommendations and agreements were effectively adhered to, that implementation was aligned with the expectations of partners, and the expectations and accountability pressures were clarified.

External communication, meanwhile, was an explicit strategy of several of the collaborative groups. Determining a collective message and communicating it beyond the collaborative served to reduce accountability tensions felt by both Forest Service staff and other collaborative members. At the same time, effective external communication can expand the collaborative group to include other stakeholder voices and perspectives. The Deschutes collaborative was particularly successful in tapping into recreational users, a group of stakeholders that previously had not been engaged in restoration-focused collaborative groups. One stakeholder outlined the strategies they deployed including:

> Outreach at the trailheads to engage groups that use the forest – snowmobilers, mountain bikers, hikers, and those kind of groups – to educate them and bring them in – actually we work with them on recommendations for the agency because we found that they have things that are important to them that we need to understand.

The success of bringing in these stakeholders led to an additional seat on the steering committee to specifically represent the interests of recreational users. This broadened the accountability system of the collaborative to include new perspectives and insights.

Fostering group education

Several collaborative groups engaged in activities to foster group education, which enhanced shared understanding, provided opportunities to clarify goals and expectations in tangible ways, and informed stakeholders about decisions. All of these activities are central to accountability, clarifying the context of decisions and answering for decisions throughout the process. These processes included strategies such as workshops, joint fact-finding exercises or joint research projects to build common understanding of the ecological context, restoration needs, effective strategies, and monitoring approaches. They also included activities such as field excursions to test assumptions and expectations on the ground. Such education efforts can resolve disagreements about scientific information while helping to clarify values.

As part of their group education processes, the SW Crown held an annual adaptive management workshop that focused on bringing new insights from research and monitoring data into deliberations about restoration strategies. The SW Jemez held a workshop on the role of fire in their ecological systems where they determined there was disagreement about the science. The Dinkey created a joint fact-finding procedure to build a common understanding of scientific and technical questions. Their educational efforts extended to developing a field guide for identifying Pacific Fisher (*Pekania pennanti*) habitats, which they used to train field crews. The Deschutes determined that some of the scientific questions they were posing had not been sufficiently addressed in the literature. They designed a research protocol to educate the entire collaborative group and the agency about ecological conditions and restoration needs in a specific system. As described by one collaborative group member:

> … this has been the mechanism by which we've introduced to the group how science and research get done – because for some people it's really like a black box – and so we've taken them out to show them how we collect the data … they really can see what goes into research design and as we discuss recommendations they can feel like they understand it better.

Resolving these differences of interpretation helped to deepen a sense of shared understanding which strengthened collaborative accountability and allowed collaborative members and the group as a whole to be answerable, and thus accountable, for activities more decisively.

Taking field excursions helps to clarify issues, serve as an education tool, allows for qualitative monitoring, builds relationships, and invites new voices to the table. One collaborator on the Deschutes summed it up, 'Field trips always seem to be that crucial ingredient that you sprinkle it throughout and good things come of them.' Stakeholders noted that field trip participants engaged in learning, trust building, and real time feedback in deliberations. One collaborative member shared that:

> Our experience has been when we get out into the field … it helps people let go of the sitting on opposite sides of the table, … more talking about 'this is what I'd like to see when I come out to the forest and this is how I would like to see my value reflected in what we do out here' … it kind of gets people away from negotiating their position and more about thinking collaboratively about what we want to see out here.

As they went into the field, changes not only emerged on the project they were reviewing, but also on later projects. Another member clarified that, 'we've seen the input from the review of one project influence the implementation of the next.'

On the Dinkey, field trips were key to reaching decisions by helping people have a sense of what worked on the ground. The field was often a location for joint education and discovery as participants evaluated forest conditions, reflected on strategies they could use, discussed priorities, and creatively thought about alternative approaches to restoration or monitoring. It deepened relational accountability, according to one Forest Service staff member, as it allowed participants to 'hear each other talk about their perspectives, talking informally, riding in the truck on a field trip, getting to know each other personally and hearing more about their situation'. At times, these trips led to a reconsideration of strategies. According to another Forest Service staff member,

> We had taken a field trip out there and this popped up, and then that became incorporated into one of the alternatives that we analysed which never would have happened if … we hadn't had them out in the field and on the ground with us.

The field also served as a place to sort through different understandings and resolve them. According to a collaborative member, '… pretty much any time you can get all the players out in the woods standing on the stump talking things over and, you know you resolve problems, little problems before they become big problems'. These trips became places to resolve conflict, address problems, deepen understanding, and develop stronger personal relationships, all of which have the potential to build greater cohesion and increase answerability across collaborative partners and agency personnel. These examples demonstrate that field trips served as an accountability measure, creating the opportunity for collaborative members to determine if agency actions were adhering to stakeholder expectations, and to allow agency staff to recalibrate their activities in relation to collaborative input.

Policy and practice implications

The creators of CFLRP sought substantive changes to how the Forest Service conducts public lands planning and management, particularly by emphasizing collaboration as an integral part of planning, implementation, and monitoring for landscape-scale restoration (Butler and Goldstein, 2010). Our research illustrates that tensions arise when collaborative and bureaucratic systems of accountability interact. At the same time, CFLRP required that the agency undertake restoration planning, implementation, and monitoring using a collaborative approach involving diverse stakeholders. Thus, to some extent, participating in a collaborative group became part of the bureaucratic and formal accountability system to which the agency is answerable. Thus, the use of strategies to strengthen accountability between the agency and collaborative groups becomes more important in a context where collaboration is mandated. We highlight several strategies that practitioners can implement to minimize tensions while maximizing accountability to multiple systems and stakeholders.

First, practitioners can focus on relationship building among collaborative participants. For the most part, these strategies facilitate stronger relational accountability, which serves as a bridge between bureaucratic and collaborative accountability, especially when tensions arise. A focus on building relationships through repeated interactions and open communication can buffer the loss of collaborative members or key agency personnel. Listening for understanding creates empathy, clarifies pressures and stress that arise from hierarchy and limited capacity, and facilitates finding commonalities upon which to build affinitive ties. Creating open forums for deliberation and education, such as field trips, joint fact-finding, and workshops can strengthen relationships as opportunities for informal and personal interaction arise more readily in venues where participants step out of a negotiation stance and into a 'what can we do together?' posture. Participants in collaborative groups and agency personnel should seek to engage in collaborative processes using these essential practices, which can help soften accountability tensions between the agency and collaborative group.

Second, collaborative accountability must also be built within the group to ensure its effective functioning and, ultimately, reaching agreements to provide recommendations to the agency. Conveners, facilitators, and participants should be involved in designing collaborative processes that can emphasize these functions to ensure collaborative accountability. Clear communication mechanisms daylight the assumptions and expectations of the group and help frame not only the agreements but also the processes used to reach them. Those frames and common understandings are grounded by group education efforts, joint fact-finding, workshops, and field trips that build scientific and technical consensus on what restoration looks like and how to achieve it. This is particularly important to the agency, which is accountable for coming up with scientifically defensible decisions; agreement on the science issues helps meet that target and softens legal risks and professional legitimacy pressures for agency staff. When consensus cannot be fully reached, minority reports allow for participants who are not in full agreement to be accountable to their constituencies and demonstrate that they pushed back against aspects of the recommendations. However, such a process also allows the agency to be accountable to the intent of the law, to receive collaborative recommendations and take them into account while adhering to planning, budget, and implementation timelines.

Third, practitioners can give attention to the agency side of the equation, thus softening some of the tensions that arise between accountability systems. While turnover is a reality, there are ways to more effectively handle turnover to ensure a smoother transition, such as integrating collaborative groups into the onboarding process and identifying replacements that are collaborative champions rather than naysayers. Bureaucratic systems will remain in place, but the system could incorporate evaluation mechanisms that include employee participation and contributions to collaborative work. In this way, agency personnel become responsible for and accountable to effective collaborative engagement. Moreover, the agency can be more transparent about decision-making processes, legal requirements, and non-negotiable aspects of projects which could go a long way toward clarifying their accountability system for other participants. Finally, capacity limitations in these cases pertained primarily to insufficient staff time. While increasing the agency's budget to hire more staff may not be feasible, clearly staff are stretched thin under increasingly complex land management contexts. Addressing budget limitations could enhance agency capacity to devote attention to restoration projects that have wide-ranging implications for reviving rural economies, reducing wildland fire risk, and enhancing ecological integrity. However, staff time may have to be addressed by increased prioritization of agency tasks. Certain projects will have to take precedence over others in order to make sure that staff can focus their time and energies on the projects that matter most.

Our cases reveal that despite significant barriers and tensions, CFLRP agency personnel and collaborative participants were able to find ways to navigate challenges that arose between different accountability systems, and, in most cases, make significant progress on getting restoration work done. Where collaboratives and agency personnel were committed to relationship building, listening, consensus decision-making with minority reporting, clear communication, and joint education including regular trips to the field, they navigated challenges and implemented innovative strategies for forest restoration. These lessons from practitioners in the field demonstrate that persistent commitment to the process and creative and thoughtful approaches to engagement can bridge the accountability divide that arises in collaborative environmental management.

References

Behn, R. (2001) *Rethinking Democratic Accountability*. Washington DC: Brookings Institution Press

Bovens, M., Gooden, R., and Schillemans, T, (Eds.) (2014) *The Oxford Handbook of Public Accountability*. Oxford: Oxford University Press

Bryson, J., Crosby, B., and Stone, M. (2015) 'Designing and Implementing Cross-Sector Collaborations: Needed and Challenging', *Public Administration Review*, 75(5): 647–663

Butler, W.H. (2013) 'Collaboration at arm's length: Navigating agency engagement in landscape-scale ecological restoration collaboratives', *Journal of Forestry*, 111(6): 395–403

Butler, W.H. and Goldstein, B.E. (2010). The US Fire Learning Network: Springing a Rigidity Trap through Multi-scalar Collaborative Networks. *Ecology and Society*, 15(3), p. 21. URL: www.ecologyandsociety.org/articles/3437.html

Charmaz, K. (2006) *Constructing Grounded Theory: A Practical Guide Through Qualitative Analysis*. Thousand Oaks, CA: SAGE Publications

Cheng, A.S. (2006) 'Build it and they will come? Mandating collaboration in public lands planning and management', *Natural Resrouces Journal*, 46: 841–858

Emerson, K. and Nabatchi, T. (2015) *Collaborative Governance Regimes*. Washington DC: Georgetown University Press

Goldsmith, S. and Eggers, W.D. (2004) *Governing by Network: The New Shape of the Public Sector*. Washington DC: Brookings Institution Press

Hahn, T. (2011) 'Self-organized governance networks for ecosystem management: who is accountable?' *Ecology and Society*, 16(2), URL: www.ecologyandsociety.org/vol.16/iss2/art18/

Hays, S.P. (1999 edition; originally published 1959). *Conservation and the Gospel of Efficiency*. Pittsburgh, PA: University of Pittsburgh Press

Innes, J.E. and Booher, D.E. (2010) *Planning with Complexity: An introduction to collaborative rationality for public policy*. New York: Routledge

Jedd, T. and Bixler, R.P. (2015) 'Accountability in Networked Governance: Learning from a case of landscape-scale forest conservation', *Environmental Policy and Governance*, 23(3): 172–187.

Kaufman, H. (1960) *The Forest Ranger: A Study in Administrative Behavior*. Baltimore: Johns Hopkins University Press

Margerum, R. (2011) *Beyond Consensus: Improving Collaborative Planning and Management*. Cambridge, MA: MIT Press

Moncrieffe, J. (2011) *Relational Accountability: Complexities of Structural Injustice*. London: Zed Books

Monroe, A. and Butler, W.H. (2016) 'Responding to a policy mandate to collaborate: Structuring collaboration in the Collaborative Forest Landscape Restoration Program', *Journal of Environmental Planning and Management*, 59(6): 1054–1072

Romzek, B. (2014) 'Accountable Public Services', in M. Boven, R. Gooden, and T. Schillemans, Eds. *The Oxford Handbook of Public Accountability*, Oxford: Oxford University Press

Romzek, B. and Dubnick, M.J. (1987) 'Accountability in the Public Sector: Lessons from the Challenger Tragedy', *Public Administration Review*, 47(3): 227–238

Romzek, B., LeRoux, K., and Blackmar, J. (2012) 'A Preliminary Theory of Informal Accountability among Network Organizational Actors', *Public Adminstration Review*, 72(3): 442–453

Romzek, B., LeRoux, K., Johnston, J., Kempf, R.J., and Piatak, J.S. (2014) 'Informal Accountability in Multisector Service Delivery Collaborations', *Journal of Public Administration Research and Theory*, 24: 813–842

Weber, E.P. (2003) *Bringing Society Back In: Grassroots Ecosystem Management, Accountability, and Sustainable Communities*. Cambridge, MA: MIT Press

Wondolleck, J., and Yaffee, S.L. (2000) *Making Collaboration Work: Lessons from Innovation in Natural Resource Management*. Washington DC: Island Press

4 Creating and sustaining collaborative capacity for forest landscape restoration

Clare M. Ryan and Lauren S. Urgenson

Introduction

This chapter presents four dimensions of collaborative capacity that Collaborative Forest Landscape Restoration Program (CFLRP) groups created as they overcame challenges in defining a shared vision of desired conditions for forest restoration. Ecosystem change in dry, fire-adapted forests requires restoration of entire landscapes. However, restoration on a large scale is difficult, particularly on United States Forest Service lands where declining budgets, complex regulations, and long standing, multiple-use conflicts have led to gridlock and litigation. Restoration decisions on Forest Service lands present complex choices among multiple jurisdictions, landowners, and communities of interests, so that the decisions are political, social, cultural, and economic, as much as they are scientific and technical (Dietz and Stern, 2008).

A key initial task for CFLRP collaborative groups is to identify a shared vision for desired conditions that serves as collective recommendations to the Forest Service for forest restoration projects. Desired conditions are the ecological, economic, and social characteristics of forest landscapes achieved through restoration (e.g., reduced risk of stand-replacing fires, maintenance of large old pines, enhanced use of timber resources, improved recreational values) and the stand and landscape structures that support those characteristics. CFLRP groups are faced with the complex and difficult task of defining a shared vision for restoration with sufficient detail to help guide management and monitoring, but also with sufficient flexibility to encompass ecological variability and scientific and social uncertainty. In order to develop and implement forest restoration strategies that support a shared vision of desired conditions, CFLRP groups and the Forest Service must build collaborative capacity to sustain those groups over time. Objectives for this chapter are to: (1) discuss four key dimensions of collaborative capacity; (2) provide examples of capacity building strategies that six CFLRP groups used as they defined shared visions of desired conditions; and (3) conclude with policy and practice implications for creating and sustaining collaborative capacity for forest landscape restoration. Our analysis draws on interviews conducted between

2013 and 2014 with stakeholders and Forest Service staff involved with six CFLRP groups in the western United States, and presents perspectives and experiences of collaborative groups over the first five years of the program.

Collaboration and the CFLRP

Restoration of fire-adapted forests is a high priority on public lands in the western United States. A management history of fire exclusion and logging for the most valuable, large trees has resulted in forests with increased vulnerability to insects, drought stress, and uncharacteristic wildfires (Allen *et al.*, 2002; Hessburg and Agee, 2003; Noss *et al.*, 2006). Climate change models predict warmer winters, reduced snowpack, and greater summer drought across the region – conditions likely to exacerbate these risks (Millar *et al.*, 2007, Spies *et al.*, 2010). Simultaneously, there is a need to stimulate the livelihoods of forest-dependent communities through the use of forest products and expansion of recreational and cultural opportunities (Franklin *et al.*, 2014). The pace and spatial extent of ecosystem change in fire-adapted forests requires restoration of entire landscapes (Franklin and Johnson, 2012; Hessburg *et al.*, 2015). However, fire-adapted forests are dynamic systems with competing social pressures and goals of management, particularly in environmentally or culturally sensitive areas, or in habitats of threatened and endangered species (Collins *et al.*, 2010; Franklin *et al.*, 2014).

Federal lands management in the west has become a lightning rod for conflict among competing interests (Wondolleck and Yaffee, 2000; Snow, 2001; Sabatier *et al.*, 2005). In response, state and federal agencies, along with non-governmental organizations, industry, tribes, and private landowners have turned to collaboration as a supplement to traditional planning and decision-making processes. Collaboration is an approach where stakeholders work together to resolve a conflict or to develop and advance a shared vision (Gray, 1989; Koontz *et al.*, 2004). By focusing on shared concerns and promoting problem solving, collaboration is touted as being effective for addressing complex resource management issues, because many issues are local, site-specific, and often cannot be easily resolved within legislatures, agencies, or the courts (O'Leary and Bingham, 2003).

The Forest Service operates under several policies that encourage or require collaborative approaches for land management, such as the Community Forest Restoration Act (2000), the Healthy Forests Restoration Act (2003), and the National Forest Management Act (1976). The Forest Landscape Restoration Act (2009) established the CFLRP, which supported an approach to forest restoration with larger spatial extents than had been feasible under previous policies and programs (Schultz *et al.*, 2012). Central tasks for CFLRP collaborative groups include identifying a shared vision of desired conditions for restoration and delivering recommendations to the Forest Service, which has ultimate responsibility for project planning and implementation on the landscape.

Collaboratively developing a shared vision for desired conditions, along with management recommendations to achieve those conditions, depends on an ongoing and functional entity (the CFLRP collaborative group), which, if it is to be successful, must develop the capacity to collaborate over time. Although collaboration is deemed both necessary and desirable as a strategy for addressing society's most difficult public problems (Bryson *et al.*, 2006), collaborative processes are not easy. Many collaborative groups struggle to get started, manage only nominal problem-solving success (Weber, 2012), and face difficulties in sustaining viability over time (Amy, 1987; Lubell, 2004; Layzer, 2008).

This chapter builds on a previous study (Urgenson *et al.* 2017a) that identified common challenges and strategies used to identify a shared vision of desired conditions. Box 4.1 displays ten common challenges identified by more than 30 per cent of interviewees across the six collaborative groups. Our analysis of the challenges and strategies the CFLRP groups used to develop desired conditions highlights several important considerations for creating and sustaining collaborative capacity in the CFLRP context.

Box 4.1 – Challenges to developing a shared vision for restoration in CFLRP cases

Different perspectives on forest landscape management and restoration objectives

- Moving from agreement on broad ideals to specific recommendations
- Different perspectives on restored stand structures
- Socio-economic desired conditions are not well developed
- Different perspectives on appropriate ecological reference conditions
- Moving from focus on ponderosa pine forest to mixed-conifer forest types

Challenges with collaborative processes

- Forest Service lacks clear collaborative decision space
- Threat of litigation
- Participatory fatigue
- Limited Forest Service capacity to collaborate
- Lack of stakeholder trust

Source: Urgenson *et al.*, 2017

Collaborative capacity

Broadly, collaborative capacity is based on the idea that complex problems require the combined abilities of citizens, governments, and non-governmental actors to be effective at solving them over the long term (Box, 1998; Bardach, 1998; Agranoff and McGuire, 2003; Weber *et al.*, 2007). Collaborative capacity has also been described as the mobilization of skilled and committed individuals and organizations, their relationships, and the resources within a given organization or community that can be leveraged to collectively solve problems (Kaplan, 2000), and the ability of individuals, organizations and groups to overcome differences and work to address mutual goals (Weber *et al.*, 2007). Capacity is often conceived at the organizational or individual level, consisting of the resources (human, financial, technical), knowledge, and processes used to achieve a goal or mission. However, in collaborative efforts multiple organizations and individuals must integrate their collective resources, knowledge, and processes to have the capacity to achieve the larger group's goals. In this context, collaborative capacity refers to the conditions needed for coalitions to promote effective collaboration and build sustainable community change (Foster-Fishman *et al.*, 2001). This requires that collaborative groups develop and sustain new organizational structures, processes, and strategies (Cheng and Sturtevant, 2012), many of which did not previously exist.

Given the significant investment of public and private resources in collaborative efforts, understanding how collaborative groups create and sustain collaborative capacity is critical for improving the effectiveness of and outcomes from initiatives such as the CFLRP. Collaborative capacity has been positively associated with long term success in problem solving (Weber *et al.*, 2007), and knowledge sharing and integration are key to building collaborative capacity (Weber and Khademian, 2008). Emerson and Nabatchi (2015) describe collaborative capacity in terms of four functional dimensions of collaboration dynamics that enable participants to accomplish their collective purposes, including: (1) resources; (2) procedural and institutional arrangements; (3) leadership; and (4) knowledge. These dimensions provide a framework for examination of dimensions of collaborative capacity that are necessary for long term and landscape level problem solving under the CFLRP.

Leveraging and sharing resources is a key dimension of collaborative capacity (Thomson and Perry, 2006; Emerson and Natabachi, 2015). Adequate funding, technical and logistical support, and skills and expertise for information gathering and analysis are examples of resources for effective collaboration (Provan and Milward, 1995; Emerson and Natabachi, 2015). Cheng and Sturtevant (2012) speak of the importance of acquiring and coordinating adequate human, technical, and financial resources needed to carry out various planning and decision-making tasks set out by the collaborative group.

Procedural and institutional arrangements include the formal and informal rules and protocols, group design, and other dimensions of a collaborative process. This aspect of collaborative capacity is indispensable in managing the repeated interactions of multiple participants over time, as well as lending legitimacy to the collaborative process (Emerson and Nabatchi, 2015). While decision-making rules and other procedural arrangements are important for collaborative processes (Ostrom, 1990; Ansell and Gash, 2008; Bingham, 2009), more formal instutional arrangements involving charters, by-laws, and memoranda of understanding may be necessary in larger and more complex collaborative efforts (Milward and Provan, 2000, 2006). Collaborative groups tend to be less hierarchical and less formal than traditional agencies, and focus on several process components, which include agreeing on the purpose of collaboration, along with structural elements such as roles, responsibilities, and decision-making authority (Bryson *et al.*, 2006). A sense of representation in decisions is crucial to obtaining broad management support (Margerum, 2011; Bartlett, 2012; Butler, 2013), and more inclusive and larger membership has been found to provide more long-term viability than a smaller membership (Margerum, 2002). In addition, groups with more diverse and balanced membership have been found to be better able to identify and prioritize goals (Koontz and Johnson, 2004).

Leadership is a critical dimension of collaborative capacity (Ansell and Gash, 2008; Bingham and O'Leary, 2008; Carlson, 2007; Cheng and Sturtevant, 2012; DuPraw *et al.*, 2012; Saint-Onge and Armstrong, 2004; Susskind and Cruikshank, 1987) and is demonstrated by initiators or champions (who get the collaborative effort started), sponsors (who provide funds, space, technology, or administrative assistance), convenors (who help motivate involvement from diverse and representative participants), facilitators or mediators (who serve as impartial managers of collaboration dynamics), participants (who represent their organizations or constituents), scientific and technical experts (who interpret complex information), and public decision makers (who support the effort and in some cases are the final decision makers) (Emerson and Nabatchi, 2015). Bryson *et al.* (2006) note the importance of strong leadership, identifying roles for formal and informal leaders and, in particular, sponsors (individuals with prestige, authority, and access to resources), and individuals who keep the collaboration going to accomplish its goals.

The knowledge dimension of collaborative capacity emphasizes the ability to effectively gather and transmit information and understanding within and across organizations (Saint-Onge and Armstrong, 2004). In addition, knowledge is a central component of adaptive resource management, where scientific and social uncertainty often compel parties to cooperate in management experiments to build understanding (Holling, 1978). Weber and Khademian (2008) note the importance of ensuring that knowledge is sent, received, and integrated to build capacity, and Emerson and Nabatchi (2015, p. 71) note that 'collaboration requires the aggregation, division, and reassembling of data and information, as well as the generation of new, shared information.' In the following sections we describe our study methods and the strategies used by CFLRP groups to enhance these dimensions of collaborative capacity.

Study methods

We selected six CFLRP groups representing a variety of geographic, biophysical, and social contexts in the western United States. All received CFLRP funding in 2010, the program's first funding cycle. To capture perspectives across a range of ecoregions we selected two cases from each of three Forest Service regions, including the Southwest Crown of the Continent Collaborative (SWCC) and the Clearwater Basin Collaborative (Clearwater) in the Northern Region (Region 1), Four Forests Restoration Initiative (4FRI) and Southwest Jemez Mountains Collaborative (Southwest Jemez) in the Southwestern Region (Region 3), and Deschutes Collaborative Forest Project (Deschutes) and Tapash Sustainable Forest Collaborative (Tapash) in the Pacific Northwest Region (Region 6).

We conducted eighty-one in-depth, semi-structured interviews in 2013 and 2014 with eighty-six respondents (four interviews involved more than one respondent), including collaborative members, local stakeholders external to the collaborative, contractors (e.g. collaborative coordinators and facilitators), and Forest Service staff associated with each project. Using a combination of chain referral (Biernacki and Waldorf, 1981) and purposive sampling (Palys, 2008), we captured a diversity of perspectives. Interviews were conducted by phone (31 per cent) or in person (69 per cent). We asked respondents to identify the primary challenges they faced in developing a shared vision of desired conditions for restoration within their collaborative, and strategies they used to overcome those challenges. Interviews were recorded, transcribed verbatim, and coded using the qualitative data analysis software Atlas.ti version 7.5.9 (Muhr and Friese 2004).

We used an iterative process of open coding to identify challenges and strategies mentioned by interview respondents. We then consolidated preliminary codes into a refined set of 'challenge' and 'strategy' labels. See Urgenson *et al.*, 2017a for a detailed description of our coding process. For this chapter, we re-analysed the 'strategy labels' to identify examples of capacity building strategies collaborative groups used as they addressed common challenges (Box 4.1) to developing a shared vision of desired conditions. Our analysis of strategy labels highlighted four functional dimensions of collaborative capacity including resources, procedural and institutional arrangements, leadership, and knowledge (Table 4.1). For each dimension of collaborative capacity, we present one or more quotes to exemplify linkages with strategies reported as particularly useful by at least one (but often multiple) respondent(s) in developing desired conditions under the CFLRP.

Findings: collaborative capacity and the CFLRP

Our analysis revealed dimensions of collaborative capacity that were important for collaborative groups as they addressed common challenges to identifying a shared vision for desired conditions. The groups used a variety of strategies including facilitation, coordination, decision rules, issue-based recommendations, sub-groups, diverse membership, mutual support, field trips, landscape-level analysis, and science engagement. Table 4.1 illustrates how the strategies used by CFLRP groups reflect dimensions of collaborative capacity.

Table 4.1 Dimensions of Collaborative capacity and related strategies used by CFLRP groups

Collaborative Capacity Dimension	*Examples of Related Strategies*
Resources	Leverage non-federal funding for planning Funding for dedicated staff (facilitator, coordinator)
Procedural and institutional arrangements	Level of support rules Issue based recommendations Sub-groups
Leadership	Diverse membership Support from collaborative Support from USFS (all levels)
Knowledge	Field trips Landscape level analysis Science engagement

Resources

CFLRP groups expanded their resource capacity by leveraging non-federal funding for planning, and acquiring other funding to support dedicated facilitator and coordinator staff positions. A key limiting resource mentioned by participants was the fact that federal CFLRP funding can be used to support restoration project implementation and monitoring, but not planning. This constraint created challenges for several groups, and they struggled to acquire sufficient resources to support their planning processes. Development of a shared vision for desired conditions is a resource-intensive, yet necessary, step in the CFLRP planning process. Funding to hire dedicated facilitators and coordinators is critical for enabling collaborative groups to establish processes, decision rules, and group norms of behaviour (which in turn influence the procedural and institutional arrangements described later in this section).

Respondents from the Deschutes and Clearwater groups described working with non-profit (501(c) 3) organizations, which enabled their groups to apply for and secure funding to hire full or part-time staff. Members of the Deschutes banded together to develop a non-profit, with the intention of supplementing their funding for key staff and activities. The Clearwater worked with a local Resource Conservation and Development Council (RC&D), which served as the comptroller for the financial resources that the group had secured. RC&Ds are not-for-profit 501(c) 3 organizations within the U.S. Department of Agriculture's Natural Resources Conservation Service (NRCS). In return, the RC&D received payments from the collaborative for their services, which provided support to continue its efforts to sustain and improve local rural livelihoods.

In several collaborative groups, one or more non-profit organizations served a bridging role as an organization with a seat at the table while also working to acquire additional funding and resources to support planning processes. For example, a Tapash member described the role of a conservation non-profit in helping fill planning resource gaps:

> We're going out there trying to find grant funding for some of the planning budgets … it's not coming as CFLRP dollars. I mean, collaboration is really about the contributions that the stakeholders can bring to the table, and that's what we're doing.

Another strategy included reliance on the willingness of some organizations to pay employees for their time spent working with the collaborative as part of their job responsibilities. These members can then serve in leadership or other roles for the collaborative. Thus, resources were provided by many different organizations who held the notion that active involvement with the collaborative group would help support representation of their interests and values at the table.

Facilitator and coordinator staff play critical roles in establishing and strengthening collaborative capacity. Facilitators are neutral parties with experience in collaboration and conflict management. They can help groups resolve conflict and move towards common ground. Several CFLRP groups contracted with third party facilitators to help members communicate and make decisions more effectively. With facilitation help, CFLRP groups used strategies that helped them make progress on setting visions, goals, and recommendations. As a 4FRI respondent reported, '… external facilitation has been crucial … to move everyone along even when there were these contentious issues. Without external facilitation we might have actually just stopped the process.'

In addition, CFLRP coordinators helped groups build capacity in terms of administrative and logistical support, which maintained the momentum of the effort, and helped address issues related to participatory fatigue. Coordinators provided administrative support and stability for collaborative groups, ensuring that the collaborative groups functioned in many other ways. CFLRP coordinator roles included meeting planning, field trip coordination, grant or report writing, science delivery, design of outreach and communication, liaison with the Forest Service, membership support, collating and summarizing stakeholder opinions, identifying issues for discussion, and organizing expert visits. As one stakeholder from the SWCC said, collaborative members were often exceedingly busy and 'prioritized' core duties of their 'day jobs', so dedicated coordinators provided valuable functions in organizing meetings, keeping collaborative members accountable, and ensuring that the group was on track. A respondent from the Tapash Collaborative mentioned that '… time is the limiting factor more than anything else, [without a designated coordinator] we would just flounder.'

Procedural and institutional arrangements

CFLRP participants described several procedural and institutional arrangement strategies that enhanced collaborative capacity, including level of support rules, issue-based recommendations, and the use of sub-groups. It is important to define these arrangements early on in the process, as described by a 4FRI respondent:

> The rules of engagement are so important to define up front. If you are going to truly be collaborative, it's probably the piece you need to work on first.... There is a lot to learn and there's a lot of discussion but it's not what catches us up usually. It's did people feel like they were part of the process and that they had an opportunity to get their piece in? If people don't feel that way, then that's where we get into problems. Ultimately, I think when collaboration breaks down, typically the things that seem to come up are sort of administrative procedure issues.

Level of support rules aided decision-making on issues where there was not full consensus. Collaborative groups often sought full consensus, but when full consensus was not possible, members with reservations developed a minority report or alternative opinion, which helped to clarify areas of agreement and disagreement, gauge support for proposed recommendations, and resolve impasses, thus allowing the group to move forward. This decision-making capacity was critical for enabling CFLRP groups to make recommendations within the agency's required timeline.

Issue-based recommendations helped groups develop capacity to scale-up agreements on desired conditions beyond individual forest stands or project areas. Issue-based recommendations were formal, written documents that conveyed the group's agreement and vision about management topics. This strategy was particularly useful in helping collaborative groups develop 'zones of agreement' that focused on specific management topics, while providing flexibility in management recommendations to the Forest Service. As described by a Deschutes respondent:

> The hope is if we've developed a recommendation you can take them from one project to the next one. They may change or get tweaked a little bit based on what we learned from applying it on this project and we might change a little bit of our recommendations.

Dividing into sub-groups, working groups, or sub-committees was another arrangement that enhanced collaborative capacity. Several CFLRPs used this strategy to focus on specific topics such as economics, restoration in different forest types, monitoring, or communications. The smaller group format allowed individuals to contribute to and work out the details on issues that mattered most to them, and was an efficient way to reduce the necessity of convening all group members, helping to tame participatory fatigue. As described by a 4FRI respondent, 'The [sub-groups] do the homework that is needed by the whole collaborative.' Sub-groups presented their findings, and the entire collaborative group was given an opportunity to discuss and give feedback. This approach boosted the larger collaborative group's capacity to move forward with decisions and take on additional issues, even though not all members were able to participate in a sub-group.

Although we highlight only a few common strategies here, it is clear that procedural and institutional arrangements are fundamental for creating and sustaining collaborative capacity. All of the CFLRP groups in our study noted the importance of attention to this dimension. And they all developed different collaborative group processes and structures to achieve their goals, which reinforces the idea that each context is different and one procedural or institutional arrangement is unlikely to work in all cases. However, each group did address the need for such arrangements in their own way and responded to emergent challenges when they arose.

Leadership

CFLRP participants identified several strategies for building leadership capacity, including diverse membership and mutual support from the collaborative group and the Forest Service. The groups varied in the extent to which they experienced challenges with Forest Service capacity and stakeholder trust, reflecting variation in local leadership and whether the group had the support of dedicated, energetic champions in both the agency and among diverse stakeholder groups.

Several CFLRP groups made explicit efforts to have diverse membership, participation, and leadership in their group. Diversity refers to broad and balanced representation of stakeholder interests in the collaborative group. Diversity improved the quality of recommendations from the group and ensured that leadership was distributed among the participants. These participants then served as 'ambassadors' to their organizations and communities, helping to cultivate broad support for projects endorsed by their collaborative group. As described by a Deschutes respondent:

> It's not as much even what the message is as who is carrying the message and that they have trust within their own circles. It's been those key people to take that message back to the communities they work with. Each of us have our own sphere of influence. So, once you have that shared vision taking that back to your sphere of influence and spreading that message.... It's taken a lot of the behind scenes work by people to go back and say, we really want to see this project to go forward.

Support from the collaborative was an important way for CFLRP groups to address the limited capacity of agency staff. The CFLRP effort placed new demands on Forest Service staff and resources, limiting their ability to take on new tasks. Some participants mentioned that a subset of Forest Service staff simply lacked interest in or the skills for collaboration. The Clearwater group made a strategic decision to work cooperatively with the Forest Service, to learn about agency timelines and processes, and to encourage community support for collaboratively endorsed projects. Mutual support between collaborative groups and the agency, and especially from local Forest Service leadership, helped overcome challenges with trust in the Forest Service and the lack of Forest Service capacity to collaborate. Forest Service support in the Clearwater helped build relationships, develop shared commitment, and fostered a culture of appreciation and respect. As one collaborator noted, 'The Forest Supervisor actually made collaborating a mandatory part of their [Forest Service staff] job.' In the Deschutes, commitment and support from the Forest Supervisor was seen as critical to the ability of managers to garner the time and resources necessary to work with the collaborative group. A respondent from the Southwest Crown of the Continent (SWCC) described the importance of support from field staff to upper level leadership:

> I feel like from the get-go that the Forest Supervisors, as best as they can, have come to meetings. Right from the start. These people have a lot of stuff on their plates and this is one part of it. Just their level of engagement in this effort, we couldn't do without it. It's great to have the Forest Supervisors there, as many meetings as they possibly can, and the District Rangers.

The CFLRP mandates collaboration, yet that is easier said than done. Leadership capacity in the CFLRP emerged in multiple ways and evolved as participant relationships developed over time. Clearly, the leadership capacity exhibited by the collaborative groups and Forest Service staff (at all levels of the agency) is necessary in order to achieve the goals of the program.

Knowledge

Key strategies for creating knowledge capacity included field trips, landscape level analysis, and science engagement. Field trips helped groups build knowledge capacity by providing specific contexts for decision-making, as well as grounding discussions in real-world examples. As described by a 4FRI respondent:

> The most helpful thing is actually going to the field and field-testing and field-evaluating. We have high levels of agreement in the field no matter who's doing it. So, we've tried to develop an implementation plan and then we've tested it in the field and we've tweaked it based on the results. And that's been important. What we were able to learn from field trips was huge.

Landscape level analysis involved a variety of tools and techniques to assess forest conditions and management options over large scales. Rather than apply a blanket management approach to their project area, landscape-level analysis enabled groups to strategically plan and place treatments to balance multiple forest uses and resource needs. To address potential trade-offs among restoration, fuels reduction, and wildlife habitat, a Tapash participant explained:

> … the other idea being that if you want to accomplish every objective on every acre you're going to struggle between fuels and owls. If you can step back and think about a landscape where you accommodate certain things on different parts of your landscape you're going to have a lot more options.

Science engagement activities included creating opportunities for shared learning about historical conditions and the implications for the restoration of contemporary fire-adapted forests (Urgenson *et al.*, 2017b) as well as involvement in research projects. Several collaborative groups solicited input from external scientists, including representatives of science-based conservation organizations, and faculty and students at local universities. Science engagement activities bolstered capacity by developing a stronger knowledge base for the collaborative group, which enhanced its ability to develop a shared vision of desired conditions. As described by one Deschutes stakeholder: 'The first step before we take a stance is to hear what research is out there and educate ourselves so that we all have at least a shared understanding of the research. Then form recommendations based on that.'

In other groups, science-based conservation organizations served an important boundary spanning role in facilitating exchanges of ideas around the contributions of ecological science, local knowledge, and social values in collaborative decision-making (see Colavito in this volume for more on boundary organizations). In several cases, developing knowledge capacity created additional transparency and facilitated greater trust among collaborative members and the agency. A Tapash respondent explained, '… the benefits are transparency, and … because it's so closely aligned with science, that helps reduce the potential for conflict and litigation'. Conversely, where knowledge products were produced without participation from the collaborative group, those products were not as readily accepted by the collaborative, as described by a SW Crown respondent, '… because it wasn't like people had input and communication through the whole process, it's not an owned product from the whole collaborative'.

Although science engagement activities helped build knowledge capacity, several groups faced challenges in their ability to access adequate ecological, and especially, social and economic, knowledge. A Southwest Jemez participant noted:

> The Forest Service has had so many challenges associated with the biological side of things that they felt like they needed to get all the biology straight … and they spent a lot of time on that and to me it's great.… But they haven't done as much on setting the desired conditions for social or economic. And so the list of social and economic are all pretty much the same generic stuff.…

Because social and economic information, in addition to ecological information, supports the primary goals of the CFLRP program, it is important to create and share strategies that will help groups build their knowledge capacity in all areas.

Policy and practice implications

The CFLRP represents an important example of institutionalizing collaboration for landscape level restoration; it requires that groups develop collaborative capacity to achieve the goals of 'collaborative, science-based ecosystem restoration of priority forest landscapes through a process that encourages ecological, economic, and social sustainability' (16 USC §7303). It is imperative to consider how CFLRP groups develop and sustain the capacity to collaborate over time, especially in light of the significant public and private resources dedicated to CFLRP. Although it was not yet clear whether each of the collaborative groups in our study had developed the capacity to sustain collaboration over a longer time period, a Forest Service assessment after the program's first five years suggests that CFLRP groups were developing a range of strategies and activities to support the program's goals of accelerating restoration, conserving forest values, and generating rural economic opportunities (USFS 2015). It is also important to note that several groups had been collaborating for a number of years prior to the creation of the program, giving them a jump-start on creating collaborative capacity.

The concept of collaborative capacity is useful for understanding how collaborative groups function to achieve their goals. Our study focused on the initial stage of the CFLRP planning process when groups worked to articulate a shared vision for desired conditions. Our findings highlight examples of strategies that can help CFLRP groups strengthen collaborative capacity in the dimensions of resources, procedural and institutional arrangements, leadership, and knowledge. More systematic study is needed to understand how different dimensions of collaborative capacity may be necessary for different stages of the collaborative process. For example, some groups may focus more on procedural and institutional arrangements early on in their process and, thereafter, may focus less as the group evolves and matures. The long-term nature of the CFLRP program, and the restoration challenges it sets out to address, provides excellent opportunities to learn, as groups move from direction setting to implementation and monitoring (Butler *et al.*, 2015).

The cases reviewed highlight an important resource constraint: federal CFLRP funds can be used for project implementation and monitoring, but not for planning activities. Planning is a resource-intensive aspect of landscape restoration, and consensus-based deliberation and shared learning to identify a shared vision require considerable time and energy. Our cases illustrate the importance of designated coordinators and facilitators who help collaborative groups create capacity in procedural and institutional leadership and knowledge areas. Coordinators and facilitators are critical to sustaining collaborative efforts over time; this requires significant investment from the participating organizations who commit staff time, funding, or both to ensure that these positions exist. To support this aspect of collaborative capacity, future policy could consider increasing the flexibility of funding uses so that coordination and facilitation as well as planning activities can be supported throughout the collaborative process.

Procedural and institutional dimensions of capacity are equally important, and one strategy shared widely across our cases suggests that complete consensus should not always be a requirement in order to move forward. Understanding and clarifying how decisions will be made and using level of support rules to reach a 'zone of agreement' enabled groups to move forward even when complete consensus was not achievable. In addition, the level of organizational formality (such as charters, membership roles, and decision-making rules) and the types of procedural and institutional arrangements necessary for any collaborative group will depend on that group's history and context (see Stern and Coleman as well as Bergemann *et al.* in this volume). Our findings suggest that regardless of the level of formality a group decides on, procedural and institutional arrangements should be explicitly considered and reconsidered as the collaborative process evolves so as to ensure that capacity can be sustained.

Leadership capacity is a complex and nuanced concept. The six CFLRP cases in our study illustrate the importance of reciprocal leadership support from the agency as well as from participants in the collaborative groups. Leadership within the Forest Service was critical to ensure that its employees were able to engage effectively with collaborative groups. Leaders within the agency can enhance collaborative capacity, not only by encouraging their employees to participate but also by maintaining a presence themselves.

Knowledge capacity often receives the most focus from the outset of collaborative efforts as participants are anxious to share perspectives and information perceived as relevant to the restoration task at hand. However, scientific information alone was not adequate to support collaborative capacity.
Even cases with an abundance of scientific information experienced challenges in decision-making and our interviews made clear that scientific information is necessary, but not sufficient, for social learning in the process of identifying and articulating desired conditions. For example, a focus on ecological information without consideration of social and economic information can alienate stakeholders. Boundary organizations played an important

role in many groups by contributing or translating information. These organizations should be recognized as key contributors to supporting and sustaining the knowledge capacity that is critical for effective collaboration (see Colavito, in this volume).

This chapter highlighted just a few examples of the strategies that CFLRP groups use to build collaborative capacity; undoubtedly there are many more examples that can be identified and shared. It would be useful for future iterations of landscape restoration policies to emphasize, if not outright require, that applicants for funding demonstrate that they have explicitly considered and planned how they will create and sustain collaborative capacity in the dimensions of resources, procedural and institutional arrangements, leadership, and knowledge. In order to share those lessons more broadly and achieve the ambitious goals of landscape level restoration, comprehensive and systematic study of all CFLRP groups is needed to enhance our understanding of how groups build collaborative capacity.

References

Agranoff, R. and McGuire, M. (2003) *Collaborative Public Management*. Washington DC: Georgetown University Press

Allen, C.D., Savage, M., Falk, D.A., Suckling, K.F., Swetnam, T.W., Schulke, T., Stacey, P.B., Morgan, P., Hoffman, M., and Klingel, J.T. (2002) 'Ecological restoration of southwestern ponderosa pine ecosystems: a broad perspective', *Ecological Applications*, 12(5): 1418–1433

Amy, D.J. (1987) *The Politics of Environmental Mediation*. NY: Columbia University Press

Ansell, C. and Gash, A. (2008) 'Collaborative governance in theory and practice', *Journal of Public Administration Research and Theory*, 18(4): 543–571

Bardach, E. (1998) *Getting Agencies to Work Together: The Practice and Theory of Managerial Craftsmanship*. Washington DC: Brookings Institution

Bartlett, G. (2012) Developing collaboration and cooperation. Pages 81–88. In M. North, (Ed.). *Managing Sierra Nevada Forests*. General Technical Report PSW-GTR-237. USDA Forest Service, Pacific Southwest Research Station, Albany, CA

Biernacki, P. and Waldorf, D. (1981) 'Snowball sampling: problems and techniques of chain referral sampling', *Sociological Methods and Research*, 10(2): 141–163

Bingham, L.B. (2009) 'Designing justice: Legal institutions and other systems for managing conflict', *Ohio State Journal on Dispute Resolution*, 24(1): 1–50

Bingham, L.B. and O'Leary, R. (Eds.) (2008) *Big Ideas in Collaborative Public Management*. M.E. Sharpe, NY

Box, R.C. (1998) *Citizen Governance: Leading American Communities into the 21st Century*. Thousand Oaks, CA: SAGE Publications

Bryson, J.M., Crosby, B., and Stone, M.M. (2006) 'The design and implementation of cross-sector collaborations: Propositions from the literature', *Public Administration Review*, December Special Issue, pp. 44–55

Butler, W.H. (2013) 'Collaboration at arm's length: navigating agency engagement in landscape-scale ecological restoration collaboratives', *Journal of Forestry*, 111(6): 395–403

Carlson, C. (2007) *A Practical Guide to Collaborative Governance*. Policy Consensus Initiative, Portland, OR

Cheng, A.S., and Sturtevant, V.E. (2012) 'A framework for assessing collaborative capacity in community-based public forest management', *Environmental Management*, 49(3): 675–689

Collins, B.M., Stephens, S.L., Moghaddas, J.J., and Battles, J. (2010) 'Challenges and approaches in planning fuel treatments across fire-excluded forested landscapes', *Journal of Forestry*, 108(1): 24–31

Dietz, T. and Stern, P.C. (Eds.) (2008) *Public Participation in Environmental Assessment and Decision Making. Panel on Public Participation in Environmental Assessment and Decision Making*. National Research Council, Washington DC

DuPraw, M.E., Cardwell, H.E., Placht, M.T., and McGonigle, T. (2012) 'Assessing the collaborative capacity of the U.S. Army Corps of Engineers: Relevant literature, a practical assessment tool, and reflections on third-party roles', *Conflict Resolution Quarterly*, 30(1): 81–132

Emerson, K. and Nabatchi, T. (2015) *Collaborative Governance Regimes*. Washington DC: Georgetown University Press

Foster-Fishman, P.G., Berkowitz, S.L., Lounsbury, D.W., Jacobson, S., and Allen, N.A. (2001) 'Building Collaborative Capacity in Community Coalitions: A Review and Integrative Framework', *American Journal of Community Psychology*, 29(2): 241–261.

Franklin, J.F., Hagmann, R.K., and Urgenson, L.S. (2014). 'Interactions between societal goals and restoration of dry forest landscapes in western North America', *Landscape Ecology*, 29(10): 1645–1655

Franklin, J.F. and Johnson, K.N. (2012). 'A restoration framework for federal forests in the Pacific Northwest', *Journal of Forestry*, 110(8): 429–439

Gray, B. (1989) *Collaborating: Finding common ground for multiparty problems*. San Francisco, CA: Jossey-Bass

Hessburg, P.F. and Agee, J.K. (2003) 'An environmental narrative of inland northwest United States forests, 1800–2000', *Forest Ecology and Management*, 178(1–2): 23–59

Hessburg, P.F., Churchill, D.J., Larson, A.J., Haugo, R.D., Miller, C., Spies, T.A., North, M.P., Povak, N.A., Belote, T., Singleton, P.H., Gaines, W.L., Keane, R.E., Aplet, G.H., Stephens, S.L., Morgan, P., Bisson, P.A., Rieman, B.E., Salter, R.B., and Reeves, G.H. (2015) 'Restoring fire-prone inland pacific landscapes: seven core principles', *Landscape Ecology*, 30(10): 1805–1835

Holling, C.S. (1978) *Adaptive Environmental Assessment and Management*. NY: John Wiley and Sons

Kaplan, A. (2000) 'Capacity building: Shifting the paradigms of practice', *Development in Practice*, 10(3): 517–526

Koontz, T.M., Steelman, T.A., Carmin, J., Korfmacher, K.S., Moseley, C., and Thomas, C.W. (2004) *Collaborative Environmental Management: What Roles for Government?* Washington DC: Resources for the Future Press

Koontz, T.M., and Johnson, E.M. (2004) 'One size does not fit all: Matching breadth of citizen participation to watershed group accomplishments' *Policy Sciences*, 37(2): 185–204

Layzer, J.A. (2008) *Natural Experiments: Ecosystem-Based Management and the Environment*. Cambridge, MA: MIT Press

Lubell, M. (2004) 'Collaborative environmental institutions: all talk and no action?' *Journal of Policy Analysis and Management*, 23(3): 549–573

Margerum, R.D. (2002) 'Collaborative planning: Building consensus and building a distinct model for practice', *Journal of Planning Education and Research*, 21(3): 237–253

Margerum, R.D. (2011) *Beyond Consensus: Improving Collaborative Planning and Management*. Cambridge, MA: MIT Press

Millar, C.I., Stephenson, N.L., and Stephens, S.L. (2007) 'Climate change and forests of the future: managing in the face of uncertainty', *Ecological Applications*, 17(8): 2145–2151

Milward, H.B., and Provan, K.G. (2000) 'How networks are governed' in *Governance and Performance: New Perspectives*. Heinrich, C.J. and Lynn, L.E. (Eds.) Washington DC: Georgetown University Press

Milward, H.B., and Provan, K.G. (2006) *A Manager's Guide to Choosing and Using Collaborative Networks*. IBM Center for the Business of Government, Washington DC

Muhr, T. and Friese, S. (2004) User's Manual for ATLAS.ti Scientific Software Development. GmbH, Berlin, Germany

Noss, R.F., Franklin, J.F., Baker, W.L., Schoennagel, T. and Moyle, P.B. (2006) 'Managing fire-prone forests in the western United States', *Frontiers in Ecology and Environment*, 4(9): 481–487

O'Leary, R. and Bingham, L.B. (Eds.) (2003) *The Promise and Performance of Environmental Conflict Resolution*. Washington DC: Resources for the Future

Ostrom, E. (1990) *Governing the Commons: The Evolution of Institutions for Collective Action*. Cambridge: Cambridge University Press

Palys, T. (2008) 'Purposive sampling' in Given, L.M., (Ed.) *The Sage Encyclopedia of Qualitative Research Methods*. (pp. 697–698). Thousand Oaks, CA: SAGE Publishing

Provan, K.G., and Milward, H.B. (1995) 'A preliminary theory of interorganizational effectiveness: a comparative study of four community mental health systems', *Administrative Science Quarterly*, 40(1): 1–33

Sabatier, P.A., Focht, W., Lubell, M., Trachtenberg, Z., Vedlitz, A., and Matlock, M. (Eds.) (2005) *Swimming Upstream: Collaborative Approaches to Watershed Management*. Cambridge, MA: MIT Press

Saint-Onge, H. and Armstrong, C. (2004) *The Conductive Organization Building Beyond Sustainability*. NY: Elsevier

Schultz, C.A., Jedd, T., and Beam, R.D. (2012) 'The collaborative forest landscape restoration program: a history and overview of the first projects', *Journal of Forestry*, 110(7): 381–391

Schultz, C.A., Coelho, D.L., and Beam, R.D. (2014) 'Design and governance of multiparty monitoring under the USDA Forest Service's collaborative forest landscape restoration program', *Journal of Forestry*, 112(2): 198–206

Snow, D. (2001) 'Coming home: An introduction to collaborative conservation', in P. Brick, D. Snow, and S. Van de Wetering (Eds.) *Across the Great Divide: Explorations in Collaborative Conservation and the American West*. (pp. 1–11). Washington DC: Island Press

Spies, T.A., Giesen, T.W., Swanson, F.J., Franklin, J.F., Lach, D., and Johnson, K.N. (2010) 'Climate change adaptation strategies for federal forests of the Pacific Northwest, USA: ecological, policy, and socioeconomic perspectives', *Landscape Ecology*, 25(8): 1185–1199

Susskind, L. and Cruikshank, J. (1987) *Breaking the Impasse: Consensual Approaches to Resolving Public Disputes*. NY: Basic Books

Thomson, A.M., and Perry, J.L. (2006) Collaboration processes: inside the black box. *Public Administration Review*, 66(1): 20–32

Urgenson, L.S., Ryan, C.M., Halpern, C.B., Bakker, J.D., Belote, R.T., Franklin, J.F., Haugo, R.D., Nelson, C.R., and Waltz, A.E.M. (2017a) 'Visions of restoration in fire-adapted landscapes: lessons from the collaborative forest landscape restoration program', *Environmental Management*, 59(2): 338–353.

Urgenson, L.S., Nelson, C.R., Haugo, R.D., Halpern, C.B., Bakker, J.D., Ryan, C.M., Waltz, A.E.M., Belote, R.T., and Alvarado, E. (2017b) 'Social perspectives on the use of reference conditions in restoration of fire-adapted forest landscapes', *Restoration Ecology*, (early on-line).

USFS (2015) *Collaborative Forest Landscape Restoration Program 5-Year Report, FY 2010–2014*. US Department of Agriculture, Washington D.C.

Walpole, E.H., Toman, E., Wilson, R.S., and Stidham, M. (2017) 'Shared visions, future challenges. A case study of three Collaborative Forest Landscape Restoration Program locations', *Ecology and Society*, 22(2): 35

Weber, E.P., Lovrich, N.P., and Gaffney, M.J. (2007) 'Assessing collaborative capacity in a multidimensional world', *Administration and Society*, 39(2): 194–220

Weber, E.P., and Khademian, A.M. (2008) 'Managing collaborative processes. Common practices, uncommon circumstances', *Administration and Society*, 40(5): 431–464

Weber, E.P. (2012) 'Unleashing the potential of collaborative governance arrangements: Getting to robust durability in the Blackfoot valley', *Journal of Sustainable Development*, 5(7): 35–47

Wondolleck, J.M., and Yaffee, S.L. (2000) *Making Collaboration Work: Lessons from Innovation in Natural Resource Management*. Washington DC: Island Press

5 Commentary on collaborative governance dynamics

Emily Jane Davis

At a recent Forest Service workshop about collaboration that I helped facilitate, participants shared some of their greatest hopes and fears. Immediately, several lamented a future where they feared they would be forced to exclusively practice something that they referred to as 'big C' collaboration. When asked to explain, they described 'big C' as formal, representative, procedure-heavy, and daunting. It would require them to engage through a defined forest collaborative group, typically with an expansive spatial scope taking in an entire ranger district or national forest. 'Little c', in contrast, was a more organic form of dialogue with local stakeholders, the composition and focus of which shifted depending on the projects, resources, and questions at hand. My co-facilitator, Meg Mitchell, and I then asked them to think beyond this dichotomy to identify principles of good collaboration. What if, we suggested, not every project or management issue could be addressed in the same fashion by the same people in one designated forest *collaborative group*? Although 'big C' might not always be just right, a meaningful commitment to doing *collaborative process* well is requisite, particularly when tools or authorities that mandate collaboration are used. Together, we developed the following list of key qualities for collaboration:

- A wide variety of stakeholders are represented, and continuously invited and kept updated, even if they do not come.
- There is interaction – talking with one another and the agency.
- Facilitation that stewards the process and maintains safety is provided.
- There is a convener, who may or may not be from the agency.
- There are agreements on the purpose of collaborating and the process, which are frequently checked.
- The decision space (as circumscribed by law, policy, regulations – and what you are deciding about) is clearly understood.
- There is a goal to seek consensus, as far as is possible among sometimes conflicting interests.

As collaboration continues to be formalized through the CFLRP and other efforts, it seems essential to reflect on both: (1) the strengths and trade-offs associated with 'big C' and 'little c'; and (2) these principles for practicing collaboration well, regardless of its form. Those of us who engage with natural resource collaboration are enduringly interested in what makes it work. The CFLRP has offered an unprecedented opportunity to pursue this favourite question anew, with numerous collaborative groups operating under similar mandates across different landscapes and contexts. It also represents the institutionalization of a particular type of collaborative governance, wherein established groups with names and spatial areas of focus form and work together beyond a single project or process. The authors of the three preceding chapters have chosen *trust*, *accountability*, and *capacity* to guide their inquiries into this CFLRP-driven collaborative approach. They bring different lenses to bear, but together affirm several common threads.

First, as collaborative groups have become institutionalized, participants have oft-mentioned, well-trodden spaces through which they travel. Most common perhaps is the meeting room with its attendant flip charts, presentations, and shuffling papers – and a neutral, skilled facilitator who sets the tone and creates safety. Another space is the field, where we hope that our abstracted understanding of concepts and differences crafted indoors will dissolve as we 'kick dirt' and observe the forest in real time. Or, perhaps, the rig, where we sit next to those with whom we disagree, talk about our children, and find that we may in fact like each other. One space that receives less coverage in academic papers, but stars in many anecdotes, is, of course, the bar visits after meetings; whisky is said to be the ultimate icebreaker. The hope for many is that travelling these spaces may help us avoid other, less pleasant ones, such as the courtroom or the expanse of a severely burned forest. These various forums for dialogue and deliberation provide a context within which collaborative groups and their participants can foster and strengthen trust, activate and deepen relational accountability, and build new capacities – particularly in relation to the practice of collaboration itself.

Second, as the authors illuminate what it takes to collaborate, they each remind us that it requires investment. Clare Ryan and Lauren Urgenson note that among the most critical procedural and institutional dimensions of capacity are facilitation and coordination, which necessitate resources; Marc Stern and Kim Coleman echo that 'resources [are] required to build trust'. Expectations of CFLRP accomplishments without dedicated funding to support such crucial aspects of collaboration may therefore be infeasible. This is a significant policy implication. However, the necessary resources are not merely financial. A collaborative group with a grant in hand must also be able to find and engage the right human capacity for their needs. This includes a skilled facilitator who fits the group well and Forest Service partners who demonstrate leadership. Beyond this, stakeholders who lead from within and contribute their time and skills are essential for true collective action. In my experience, providing technical assistance and reviewing grant proposals, I see groups often seeking a type of 'magic facilitator' who will possess myriad traits of mediator, coordinator, and executive director in one person, calling participants to a meeting once a month and handling much of the rest. In these calls for resources, we must also recognize that further strategic understanding and investment in the collective capacity of all collaborative group stakeholders and agency partners is merited.

Third, these chapters show how collaborative groups work in consistently dynamic environments. There are inevitable turnovers, transitions, and uncertainties in people, ecosystems, and political direction. Stern and Coleman suggest that trust 'buffers collaboratives against internal or external shocks', and that 'healthy stores of diverse forms of trust' must be built. To overcome turnover, Christensen and Butler call attention to the need for relationships grounded in trust, and transparent and strong communication. Instability in the collaborative context is now well-recognized and is not likely to change. Therefore, rather than merely reiterating this familiar challenge we must seek means to address it, and make themes of resilience and adaptation more central to how we study and practice federal forest collaboration.

These discussions are relevant for all communities working on the management of collective resources to achieve ecological and social sustainability. Ample research in areas such as organizational development, public administration, and climate change shows that collaborative groups are not the first form of environmental governance to experience change and challenge, and offers insights that we should explore more deeply. In other countries with even more turmoil and transition than the United States, how does civil society reorganize to adapt? What does it mean for a government and its stakeholders to adaptively co-manage resources in a meaningful fashion, particularly when much power remains with that government?

Perhaps most important in these chapters is when the authors go beyond operational matters of 'what it takes' to succeed and persist, and remind us of other, larger questions about the context in which federal forest collaboration sits. It is not solely about how people in a room might reach agreement about management actions or come to like each other despite their differences. These types of social outcomes, while important, do not automatically or necessarily result in others, such as increased forest restoration or economic activity. Much relies upon the Forest Service itself, which possesses the ultimate decision-making power and implementation capacity on national forests. As Christensen and Butler point out, 'Rarely does the collaborative process change the underlying system of accountability, authority, and responsibility for managing public lands. Instead, it becomes an added layer with the potential to influence decisions and actions of responsible parties.' Some critique collaboration for this very possibility, claiming it allows various interests to exercise undue power over the Forest Service. Both industry and environmental groups have argued that the other side dominates collaborative discourse, while others in some rural communities see collaboration as a vehicle for restricting public access and use of national forestland. Still, others argue that collaboration in fact has minimal, if any, effect on agency decisions except as a means to legitimize them (Wynsma, 2014; Blue Mountains Biodiversity Restoration Project, 2015).

These many slings and arrows flung at collaborative groups simultaneously ascribe to them an impossible character of both excessive power and no power. They indicate the inextricable tension between formality and informality that exists; collaboration at its core is an intangible, social process that relies heavily on the dependent variables of its people and places, yet it has become institutionalized as a governance approach expected to deliver tangible outcomes. Identifying as a collaborative can allow access to resources and benefits. I have fielded inquiries over the years asking, 'what must we do to qualify as a real collaborative?' – in other words, to embody 'big C' collaboration? The clamour to 'certify' or vet collaboratives with mandatory requirements has grown. Some programs may require collaboratives to possess certain qualities to apply but, by and large, there are no steadfast arbiters or specific gates of entry.

What does this all imply for the future of federal forest collaboration? Does it suggest that we must practice 'big C' to achieve the many outcomes that this form of governance promises? The institutionalization of groups can foster their stability in some ways. Other research (Cheng and Sturtevant, 2012) explains how more well-organized, established groups may be best equipped to attract and mobilize resources. Yet this also poses challenges. For example, new groups may be pressed to gain ground and become competitive for grants or other opportunities if the 'winners keep winning', and there are limited resources to go around. Moreover, we may also collectively run the risk of entrenching ineffective organizations and inhibiting fluidity and adaptation. Groups that are longstanding are not automatically or necessarily successful. In some cases, they may be able to pass through the certification processes, checking off boxes of assumed best practices on paper, yet they may lack true energy, interactive dialogue.

As I finished reading these three chapters, I found myself revisiting these substantial questions. I also reflected on the notion of change, often viewed in forest collaboration through the prism of detrimental turnover and loss. What if we viewed change not only as a roadblock, but also as a potential source of novelty and a signal that reorganization of collaborative governance may be in order? If stakeholders and their agency partners are able to deliberately rearrange how they work together in these dynamic environments, and at varying spatial scales, they may become more adaptive and effective. The precise organization of each collaborative process or group would not be durably fixed or mandated, and people and interests would continue to inevitably shift as they do. We will not always have the luxury of seeing the same faces each time we are in the forest, the rig, or the bar – and we may not always be in these spaces under the auspices of the established collaborative group. But, perhaps, the principles for good collaborative processes may endure. Rather than focusing on the form collaboration takes, we might pursue questions of its quality and how to best achieve it in the ever-changing context of federal land management.

It seems likely that if we do this, we will still see that the well-proven findings remain true: factors such as quality facilitation, time in the field, and dedicated resources matter – and grantors, agency partners, and others will increasingly invest in those groups that possess them. Hopefully, we will also continue to learn how those factors not only create and sustain groups, but also support meaningful, adaptable dialogue about federal forest management. The three preceding chapters move beyond defining effective collaborative practice by explaining how certain characteristics and behaviours in groups and individuals can foster these necessary conditions: multiple forms of trust; broadened accountability; and durable collaborative capacity. In doing so, they offer the start of a timely roadmap for future research, practice, and policy.

References

Blue Mountains Biodiversity Restoration Project. (2015). "Collective Statement on Collaborative Group Trends." Retrieved 1 November 2017, (https://bluemountainsbiodiversi-typroject.org/collective-statement-on-collaborative-group-trends/)

Cheng, A.S. and Sturtevant, V.E. (2012). A framework for assessing collaborative capacity in community-based public forest management. *Environmental Management*, 49(3): 675–689

Wynsma, B. (2014). 'U.S. Forest Service Collaboration Process: Solution or Sham?' Retrieved October 25, 2017 (www.evergreenmagazine.com/u-s-forest-service-collaboration-process-solution-sham/)

Part II

Science and adaptive management in collaborative restoration

6 From conflict to shared visions

Science, learning, and developing common ground

Eric Toman, Emily H. Walpole, and Alexander Heeren

Introduction

As described in other chapters in this volume, public natural resource management in the United States has long been a contested endeavour (see Schultz and Butler in this volume). Natural resource management agencies were founded on the belief that science-based management would provide a means to rationally manage resources to ensure that they met the needs of current and future generations. However, public natural resource management has grown increasingly complex. Ecological systems are marked by a diversity of biophysical components that interact in a multitude of ways. In some cases, management actions that modify components or processes within these systems have resulted in unexpected outcomes; for example, in the western United States decades of fire suppression has resulted in the accumulation of fuels increasing the severity of wildfires. Adding to this ecological complexity, natural resource management decisions involve competing beliefs about the desired outcomes and appropriate uses of landscapes. Ultimately, such conflicts do not have definitive solutions; instead, preferred approaches are influenced by a range of psychological (e.g. values, beliefs) and social (e.g. trust in the decision maker, group affiliation) factors.

Ecological restoration has been defined as a 'process of assisting the recovery of an ecosystem that has been degraded, damaged, or destroyed' (Society for Ecological Restoration, 2002). Restoration goals are typically determined by identifying ecological conditions at an historical reference point prior to degradation. However, over time, ecological restoration has been increasingly recognized as an expansive concept with close links to cultural and social factors (Martin, 2017). Indeed, a range of psychological and social factors are embedded in the act of determining that certain conditions are more desirable than others, and that they merit intervention to restore (Gobster and Hull, 2000; Vining *et al.*, 2000; Egan *et al.*; 2011). Ultimately, in addition to providing ecological outcomes, restoration efforts seek to provide societal benefits (Gobster and Hull, 2000; Society for Ecological Restoration, 2004; Martin, 2017). Legislation establishing the CFLRP explicitly emphasized provision of community and economic benefits linked to restoration actions (16 USC §7303).

While the general idea of ecosystem restoration is likely to be acceptable to a diverse range of stakeholders, conflict may emerge when transitioning from the conceptual level to implementing restoration treatments on the ground. Indeed, to agree with the need for restoration, participants must first agree with the premise that the current conditions of an ecosystem are undesirable. Such agreement is not a given. For example, a proposed restoration project in Chicago encountered substantial public resistance as local residents preferred the current, non-native forest to the native prairie conditions (Barro and Bright, 1998; Vining *et al.*, 2000). Conflict may arise due to disagreements about the historical reference point used to describe target conditions, how to determine when a departure from a reference point merits action, and how restoration can be balanced with current human uses and a changing ecological and social context (Vining *et al.*, 2000; Egan *et al.*, 2011; Thompson *et al.*, 2009). Recognizing that restoration is both an ecological and social endeavour emphasizes the importance of developing a shared vision of the future state of the landscape and the processes by which those conditions will be achieved.

Using a mental models approach and thematic analysis, we considered how participants in three CFLRP locations conceptualized ecological restoration, the alignment of these models with one based on the current scientific literature, and the role of the collaborative process in advancing restoration objectives. Limited research has investigated how decision makers and stakeholders understand ecological restoration, even though such understanding plays an important role in shaping goals, beliefs about appropriate treatments, and, ultimately, the potential for conflict between participants. We address this gap through interviews with stakeholders and decision makers engaged in CFLRP restoration efforts.

Study sites and participants

This research was conducted at three CFLRP locations: the Missouri Pine-Oak Woodlands Restoration Project (Mark Twain National Forest, MO); the Southern Blues Restoration Coalition (Malheur National Forest, OR); and the Tapash Sustainable Forest Collaborative (Okanogan-Wenatchee National Forest, WA). Study locations were selected so as to provide diversity in social/psychological factors that may influence restoration visions and different collaborative structures on achieving restoration success. At all three study locations, decreasing resilience to disturbance events, departure from historic ecological conditions, and opportunities to contribute to local economic needs were cited as an underlying rationale to engage in forest restoration activities.

We conducted a total of twenty-five interviews across our three study locations. We selected participants to represent the range of agency employees, NGO members and industry members involved in restoration decisions (see Table 6.1 for description of participants) to obtain a variety of

Table 6.1 Study participants

Position	*Pine Oak*	*Southern Blues*	*Tapash*	*Total*
Forest Service line officer	2	2	2	6
Forest Service resource specialist	4	4	4	12
Conservation NGO	2	2	1	5
Industry representative	1	1	–	2
Total	9	9	7	25

perspectives. A majority (72 per cent) of our participants worked for the US Forest Service, including both line officers and resource specialists. Key non-Forest Service stakeholders, including the timber industry and conservation group representatives, were also included when actively involved in the collaborative group in each location.

Mental models

A mental models approach is one of several conceptual mapping techniques that can be used to develop a visual representation of how a person thinks about a concept, issue, or situation (Byrch *et al.*, 2007). Each mental model is unique to an individual and is comprised of the beliefs, knowledge, and experience that the individual draws on when considering a particular topic (Jones *et al.*, 2011). Individual mental models can be compared and then compiled to create a generalized model across a range of participants. Per the methodology outlined by Morgan *et al.* (2002), this research was completed in two phases. First, the research team developed a technical model to represent the current scientific understanding of forest restoration based on the scientific literature. Second, the research team conducted semi-structured interviews to probe participant conceptualizations and develop participant models. To begin, the research team systematically reviewed forty-three peer-reviewed articles and books in forest and restoration ecology to identify key concepts and relationships described in the literature. This resulted in an initial technical model that included over 150 concepts. We then shared the influence diagram with five scientists with expertise in restoration ecology (including ecologists and social scientists) who reviewed the included concepts for accuracy and completeness, and also provided guidance on how the presented concepts could be condensed. Based on their feedback, a condensed technical model was created that grouped the most relevant concepts into four main content areas (described briefly in Box 6.1 and visually represented in Figure 6.1).

We then developed an interview protocol informed by the technical model with questions organized around each of the four main content areas. We began with open-ended questions to allow participants to independently raise concepts they associated with restoration and to allow the emerging

Box 6.1 – Technical model content areas

The four main content areas in the technical model included:

- **Natural or Reference Forest Ecosystem:** Historic or natural species composition, structure, and functions. The historic disturbance regime and role of disturbance in the historic forest. The historic reference point being used and how it was selected. The landscape context of the historic condition.
- **Current Condition:** Current species composition, structure, functions, and disturbances. The attributes and causes of departure from the historic or reference condition. Land use patterns that contributed to the current condition. The landscape context of the current condition.
- **Restoration Stages:** Defining desired future conditions (including ecological, social, and economic goals) and intermediate conditions. Developing implementation methods including selecting management actions and incorporating social influences and public input.
- **Social Influences:** The role of past conflict or trust between stakeholder groups on restoration planning or activities. The influence of public perceptions of current forest conditions, historic forest conditions, or forest management more generally on the restoration effort. Social dynamics within agencies, between agencies, with the general public, and with external stakeholders. The role of collaboration and creating a shared vision for the restoration project. Economic and policy influences on restoration planning and activities.

mental model to reflect their thinking. If concepts from the technical model were not raised in response to the open-ended questions, the research team followed up with specific questions related to key concepts within each content area. The interviews were recorded and transcribed so we could code the transcripts – to distinguish which concepts were freely brought up by participants or given in response to open-ended questions versus those that emerged after specific follow-up questions. This approach allowed us to assign a degree of salience to the topics for each participant (i.e. topics offered spontaneously or after follow-up probes as well as frequency of discussion).

Once developed, these individual models were compared using a quantitative accounting to identify the presence, absence, and salience for each topic in the technical model for each participant. Moreover, these individual mental models were compiled and compared to the technical model to assess alignment with current scientific understanding (Morgan *et al.*, 2002; Jones *et al.*, 2011).

Thematic analysis – collaboration

In addition to the mental model comparisons, we completed a thematic analysis of interviews following the responsive interviewing approach described by Rubin and Rubin (2005). All interviews were transcribed verbatim and uploaded into MaxQDA V11 software (developed by VERBI) for coding. Interviews were reviewed and initially coded, following concepts included in the technical model. The research team then reviewed memos written during the interview, transcription, and coding process to identify concepts or themes raised as important – across interviews associated with the perceived benefits of and barriers to working collaboratively to achieve restoration objectives. We then conducted line-by-line thematic coding of interviews focusing on the participants' descriptions of their experiences, with collaboration as part of the restoration process. Through this process of memo review and open coding, a second coding structure was developed, related to the identified successes and challenges associated with collaboration at these three sites.

A subset of our interviews was coded by two members of the research team. Codes were then compared and discussed, and clarifications added to the code manual until an acceptable inter-coder reliability rating (Cohen's Kappa > 0.8) was achieved (Altman, 1991).

Results

Comparison of technical and participant models

The technical model of forest restoration is presented in Figure 6.1 (some subcodes and linkages were excluded from the Figure to improve readability). Codes are organized around the four main content areas (in black).

Participant models (Figure 6.2) were found to be strongly aligned with the technical model. Indeed, all of our participants exhibited high levels of agreement with most areas of the technical model; when prompted, participants were able to describe most areas of the technical model accurately, and responses conveying a true lack of knowledge (i.e. responding 'I don't know' to a prompt) were quite rare. Only a few codes were present in the participant models that were not in the technical model; these codes were generally specific to both the CFLRP as a program and the partnerships at each site. These findings suggest participants generally view ecological restoration as being consistent with the current literature, particularly in an ecological and technical sense (i.e. perceptions of historic and current conditions, and management implementation). Moreover, there was substantial overlap in participant mental models within and between locations. These findings suggest a shared conceptualization of many aspects of ecological restoration among participants – an important foundation for contributing to success of collaborative efforts (Gray, 1989; Wondolleck and Yaffee, 2000; Margerum, 2011) and discussed in more detail below.

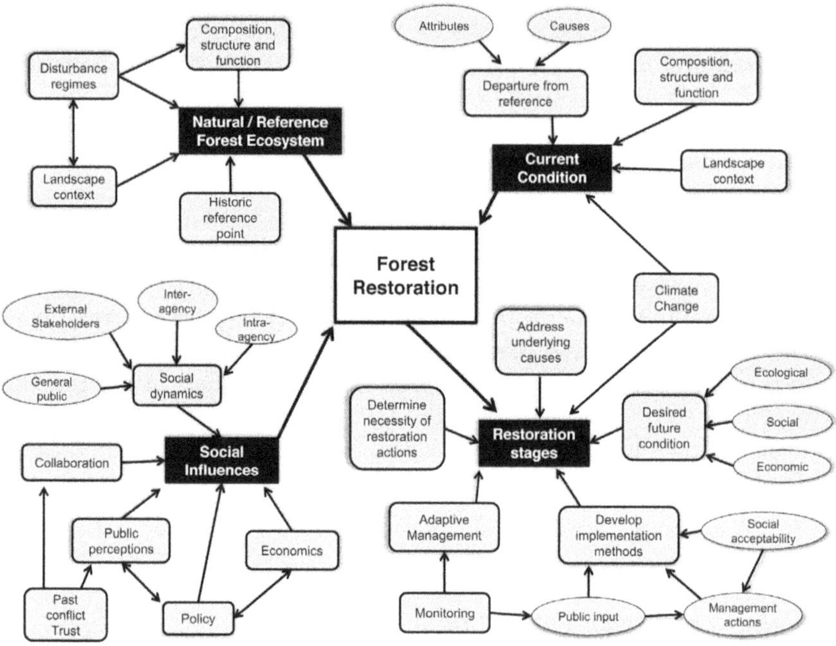

Figure 6.1 Abbreviated technical model of forest restoration developed from literature review. Four content areas are illustrated in black boxes with associated codes and sub-codes in grey.

Note
Some codes and linkages between topics are not displayed for presentation purposes.

Ecological factors

Generally, there was substantial agreement between technical and participant models in the ecologically based content areas. Most participants described all major aspects of historic forest conditions in substantial detail without prompting. Moreover, participants generally provided consistent descriptions of the historic ecological conditions used to guide restoration plans within their region.

Participants also provided detailed and consistent descriptions of the current species composition, structures, functions, and disturbances that were accurate to their location. They identified and discussed the attributes and causes of departure from the historic or reference condition without prompting, and were able to cite specific causes of degradation, including, most commonly, fire suppression, followed by selective harvesting practices, grazing, and mining. Nearly all participants described current conditions as having departed from natural conditions and being undesirable, and generally

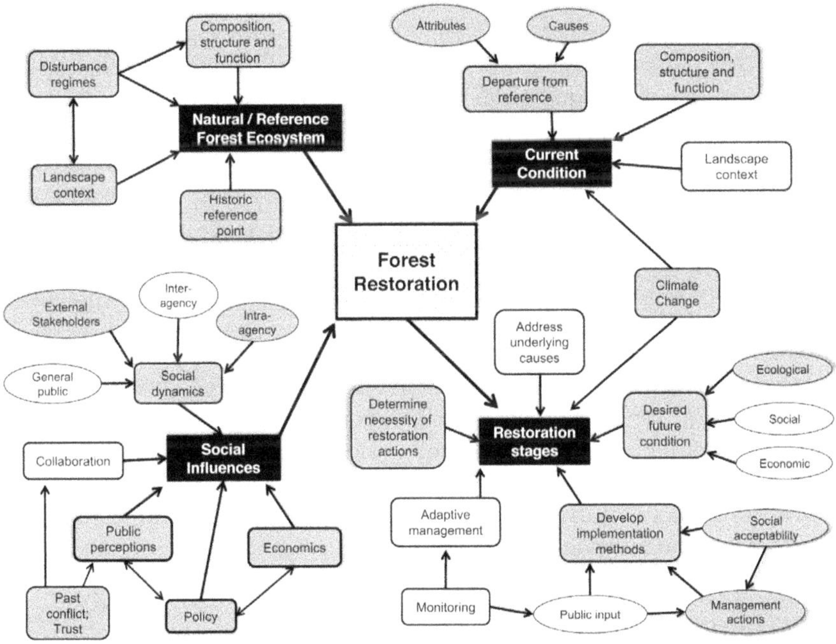

Figure 6.2 Salience of concepts for study participants. The four content areas are illustrated in black boxes. Concepts associated with each content area are shaded according to salience among participants. Dark grey: high salience – codes were raised unprompted by more than half of all participants. White: low salience – less than half of participants provided unprompted comments on this topic.

associated this departure with a perceived need for management intervention to move the system toward historic conditions. These results suggest agreement among participants on the need for restoration and general reference points in order to proceed with restoration activities.

Social factors

While overall alignment was high between the technical and participant models, some areas appeared less salient to participants. The topics associated with 'social influences' and some social aspects associated with 'restoration stages' appeared less salient. For example, while participants readily discussed the influence of policy, economics, and public perceptions on restoration efforts – typically characterizing these as constraints on achieving restoration goals – many of the other topics in the 'social influences' content area were mentioned unprompted by less than half of participants. Although the

CFLRP program has an explicit focus on collaboration, 'between-agency dynamics' and 'collaboration' were raised unprompted by fewer than half of our participants. A similarly low number of participants spontaneously discussed relationships and interactions with the general public, although social dynamics with external stakeholders and within agencies appeared to be more salient concepts. In a similar pattern, while social acceptability was commonly discussed by participants, typically as a barrier to using preferred management actions, public input processes associated with development and implementation of restoration plans were only raised by most participants after being prompted. Last, while every participant readily described ecological aspects of desired future conditions, fewer than half mentioned social or economic goals without prompting.

These findings suggest that while many of our participants recognized that social factors had an influence on their ability to achieve their restoration goals and social dynamics within their own agency and external stakeholders, they appeared to be thinking less about social and economic goals and general public relationships and involvement, or the process of collaboration itself. This finding is particularly striking given the explicit focus of the CFLRP on collaboration and providing social and economic benefits.

Between-site comparisons

Overall there was substantial consistency between sites within ecological content areas (e.g. 'natural or reference system' and 'current conditions') and technical topics (e.g. the 'process of restoration' and 'management actions'). However, some notable differences emerged regarding social influences on restoration. Overall, participants in the Southern Blues made substantially more unprompted comments regarding collaboration, working with external stakeholders, and how public input was solicited, compared to those in the Pine Oak or Tapash projects. Indeed, the influence and process of 'collaboration' was not discussed unprompted by any participants at Tapash, while 22 per cent of the participants at Pine Oak, and 78 per cent of the participants at Southern Blues mentioned 'collaboration' without prompting. In addition, a higher proportion of Southern Blues participants discussed, unprompted, economic and social goals of restoration efforts compared to participants at the other two sites, making them more closely balanced with the salience of ecological goals. However, a higher proportion of participants from Tapash discussed social dynamics between agencies in negotiating restoration efforts compared to the participants at the other locations – as well as the presence of incompatible goals and trade-offs. This suggests that there are distinct salience patterns regarding the role of collaboration and collaborative partners, social goals, and public input between locations that may reflect the amount of focus given to these activities at each site. Results from the thematic analysis explore these differences in greater detail below.

Interview themes – collaboration

As discussed elsewhere in this volume, natural resource management, particularly on public lands in the United States, has long been characterized by conflict as stakeholders compete to advance their preferred visions for future conditions and prioritize their preferred uses of the landscape. This conflict has complicated efforts for public natural resource agencies and has resulted in appeals and litigation challenging decisions and delaying or blocking management activities (Keele *et al.*, 2006). Our study locations have felt the effects of such conflicts – particularly our two locations in the Pacific Northwest, Tapash, and Southern Blues, each of which has a history of divisive disagreements over previous management activities. In this section, we explore contributions from our findings regarding how using a collaborative process may help facilitate ecological restoration efforts, and we consider whether there are remaining barriers to long-term success. Additional details regarding findings from this analysis can be reviewed in Walpole *et al.* (2017).

Shared vision, common goals

At our three locations, ecological restoration seemed to provide an effective framework to bring people with potentially disparate views together. At the conceptual level, there was general support for restoring forest conditions and, in each location, participants indicated the collaborative approach had contributed to reaching agreement on restoration goals and identifying shared challenges. Participants described a transition from agreement on general goals toward more specific questions of how restoration outcomes could be achieved on the ground. Ultimately, success at moving from the planning stage to implementation of specific projects varied between groups and by topic.

In the Southern Blues, participants described an emphasis early on in their efforts to develop a shared vision for restoration outcomes. Although this area had been characterized by a history of conflict over natural resource management decisions, as illustrated in the following quote participants were able to find agreement with the need to improve forest resilience:

> Some of the meetings were heated, some people stormed off and never came back, some stormed off and came back.... But once you get out there and start talking, you get your shared goals that the forest is not the way it should be and we have to really make it better somehow. You have a shared vision.

Participants in the Southern Blues indicated that the focus on restoration and enhanced resilience provided an effective point of agreement to bring together a diverse group of stakeholders that had not previously recognized much common ground. Participants noted some successes progressing from initial conceptual agreement to more specific topics regarding the need for management interventions to achieve desired future conditions, including ecosystem types and using treatment approaches (e.g. timber harvesting) which had been heavily contested in the past. Participants also discussed how collaboration supported the value of the local forest industry while achieving restoration goals.

Participants at Tapash described the concept of 'shared visions' less frequently, but noted there was agreement about the importance of a 'healthy forest'. They emphasized the use of a specific modelling program that allowed them to test potential outcomes of different treatment approaches and areas of implementation, and described the value of this program in helping reach agreement on specific projects.

At Pine Oak, participants described a long-standing, shared goal of restoring pine-oak woodlands that pre-dated the initiation of the CFLRP. Like the other locations, the concept of restoration proved an effective framework around which they were able to develop a shared vision and recognize areas of common ground. Individual projects were then considered based on their contributions to this shared goal.

Across locations, participants also described how a collaborative focus on restoration had provided a framework for decision makers and stakeholders with diverse views to reach agreement on a vision for desired future conditions – often focused on improved forest resilience and a shared conceptualization of the current problem (e.g. changes in forest structure and increased risk of uncharacteristic and severe fire events). Moreover, this shared vision provided a foundation that generated discussion on how such resilience could be achieved, including intervention in some areas or the use of management actions that may have been contested in the past, such as conducting prescribed burns, implementing thinning treatments, or closing roads.

Improved relationships and trust

In all three locations, participants indicated that collaborative efforts had contributed to improving relationships and trust. This was particularly evident in the Southern Blues, which had an extensive history of conflict that typically pitted industry against environmental groups in disputes about management plans on the Malheur National Forest and adjacent public lands. Participants here described how the emphasis on ecological restoration encouraged them to extend their thinking beyond a short-term focus on achieving their individual objectives to considering long-term outcomes across the forest. Focusing on improving resilience also helped group members identify areas of common ground they had not previously recognized between themselves and others (e.g. the recognition that no one wanted severe fires and that active management is required to prevent such fires).

Although restoration provided a relevant focus to bring previously disparate groups together, participants indicated that a concerted effort was still required to gradually improve relationships and to rebuild trust that had been strained due to previous interactions. They particularly mentioned the importance of intensive mediation and the value of interactions in the field and informal social contexts. One participant from the Southern Blues stated:

> When you sit in a Forest Service office, it's too clinical in a sense. Everybody can sit in their position and everybody's just kind of hanging on to their place.... So, when you get out in the woods and you walk around and you have casual conversations and you visit with people more, and then maybe after hours you go have a few beers every so often, agreements start to be forged, slowly.... If I like you and I respect you as a person, it's gonna be easier [for me to agree with you].

Such efforts take time; some participants noted spending several months engaged in relationship-building activities before discussing specific projects.

Sense of progress in management

While the full extent of the outcomes of these collaborative efforts will take time to become apparent, participants noted signs of progress. Speaking generally, participants in the Southern Blues and Pine Oak described a sense that 'things were changing' from the often adversarial approach that characterized past natural resource management decisions in their area. To be clear, participants were realistic about these ongoing changes and did not suggest that there would be a complete lack of conflict in the future. Indeed, some indicated the efforts to date were largely a result of focusing on areas of high agreement and would be tested when addressing more controversial issues.

Participants in the Southern Blues and Pine Oak reported that investing in collaborative interactions had allowed them to implement restoration treatments on a greater number of acres than they felt would have been possible otherwise. In the Southern Blues, participants noted that they were ahead of schedule in their proposed treatments and were seeking to expand the geographic area included within their designated management area. In Pine Oak, participants indicated that the shared vision for restoration and strong relationships had facilitated coordinated treatments across property boundaries and that other resource management agencies, and even some private landowners had coordinated efforts to extend treatment areas.

Participants in the Southern Blues also noted progress which they attributed to the collaborative efforts involving timber harvesting, a topic with a substantial history of conflict within the local area. As a local timber mill considered shutting down, the collaborative group engaged in discussions about how the mill could be saved with support of the conservation interests within the collaborative group. This was described by one participant as follows:

> ... It was interesting when the mill threatened to shut down about a year ago now, the last mill in town, the first person I heard saying 'we can't

> let that happen, we need the mill to do the forest restoration' was an environmentalist.

Subsequently, the group successfully negotiated an amendment to their Forest Plan, allowing the removal of larger diameter trees. In past forest planning efforts, the agency had agreed to limit the harvest of trees to those twenty-one inch in diameter at breast height (dbh) or less; however, the collaborative group helped the agency work toward eliminating the diameter cap in favour a scientific method developed by Drs Robert Van Pelt, Jerry Franklin, and Norm Johnson for identifying and protecting old growth trees, which better met the restoration goals. This change was recognized as an indication of trust in the local managers; larger trees would not be removed solely for economic gain but, if removed, this would be done to achieve shared restoration goals. These efforts were recognized by group members as strengthening their relationships; they were able to put aside past differences and work together to achieve a commonly recognized goal. As one participant pointed out:

> ... that's why we're in accelerated restoration now, because it had the collaborative behind it. It wasn't just the loggers saying 'we need timber to survive'... when you have these other voices saying that, all of a sudden I think the politicians, the Regional Foresters ... [it] really got their attention.

In addition, interviews with US Forest Service personnel in both the Southern Blues and Pine Oak suggested that the collaborative efforts also had a positive influence on the morale of agency personnel. In the Southern Blues, agency participants used terms like 'amazing', 'exciting', and 'a watershed event' to describe the success they had experienced through the collaborative effort. At Pine Oak, several Forest Service employees conveyed an intense personal investment and pride in their restoration-related work and an expectation of the legacy they would leave behind.

Remaining challenges

Although participants in all three locations were generally positive about the successes gained through the collaborative focus on restoration, they also identified several challenges that the groups had not yet addressed. One primary theme to emerge from these comments was a remaining dispute about where restoration treatments were needed and questions about how substantial a departure from historic conditions was required to warrant intervention. More specifically, in both the Tapash and Southern Blues projects, participants disagreed about the need for treatments in mixed conifer and old growth forests as well as in riparian areas. These differences seemed to reflect pre-existing divisions and historical conflict regarding management in these systems. Although participants had developed a shared vision for restoration that included support for active management intervention in more departed or high-risk systems (e.g. dense pine stands in dry areas), there were some limitations to this agreement. Moreover, participants also noted a lack of trust related to past conflicts that made reaching agreement on management in these systems particularly sensitive.

There was also debate about how economic and ecological considerations should be balanced when developing restoration plans. Participants expressed a broad range of beliefs on this topic, from those who felt it was critical to include profitable materials (generally larger diameter trees or other merchantable timber) when planning treatments to offset costs and support local communities, to others who felt ecological considerations should be strictly prioritized.

Policy and practice implications

Several implications emerge from this research for practitioners engaged in restoration or collaborative efforts. Our findings suggest that a shared vision about restoration and future forest conditions can provide a meaningful foundation to reach agreement on management actions, even among participants who had previously experienced substantial conflict. This is likely influenced, at least in part, by general agreement among stakeholders with diverse beliefs about natural resource management that restoration is a worthwhile goal. In each of our study locations, we found substantial alignment between how participants conceptualized forest restoration and the technical model developed from existing literature. This conceptualization was generally consistent across professional disciplines, agencies, and external stakeholders, and likely contributed to the shared conceptualization of ecological restoration among participants. Overall, the process of developing a shared vision was cited as critical to helping participants recognize common ground with other stakeholders, including some with a history of conflict over resource management decisions.

Based on these findings, we emphasize the importance of developing a shared vision of future conditions in the initial stages of collaborative efforts to provide a foundation for continued work. In some cases, this agreement may be readily apparent when a collaborative group is initially convened. In others, it may be necessary to engage in efforts such as joint education or fact-finding to help build this common understanding. Our participants described a substantial investment of time in meetings, field visits, and informal discussions in social settings to develop this shared vision and build relationships among participants.

A related implication is that conceptual mapping techniques, such as the mental models approach used here, may be a tool to help mediate collaborative processes. Conceptual mapping can be used to develop a visual representation of how a person thinks about a concept, issue, or situation (Byrch *et al.*, 2007). A variety of methods can be used to develop conceptual maps, varying in their level of complexity and specific approach. However, each is designed to help participants sort through and articulate their own thinking about a topic. This process can help participants clarify their own understanding of an issue including recognizing gaps or inconsistencies in one's own understanding of a system (Ackermann *et al.*, 1992). Sharing conceptual maps with a broader collaborative group can also provide a focusing tool for deliberation as participants can see where overlap may exist between different stakeholders, identify gaps in one's own understanding of a system, and illustrate the complex nature and multiple interacting variables associated with natural resource management decisions (Toman *et al.*, 2015).

It is important to note that establishing a shared vision about future conditions at a general level will not necessarily result in agreement on specific management approaches, or the need for action in all locations. While our participants had made considerable progress towards reaching agreement and overcoming issues with historical tensions in clearly departed systems, there was still substantial disagreement about the need for management interventions in some other systems. Progress in highly contested systems, or those where the stakes may be viewed as particularly high (e.g. old growth forests in our two PNW locations), will likely require additional time to develop the necessary trust among participants and agreement on needed actions. As found in previous research, beginning management efforts with projects with a limited geographic scope (Moote and Becker, 2003; Antuma *et al.*, 2014), or in areas of high agreement among stakeholders (Wondolleck and Yaffee, 2000; Moote and Lowe, 2008; DuPraw, 2014) can contribute to building relationships, trust, and a momentum that enables groups to address more contentious topics in the future. While our groups had not yet tackled the need for treatments in some contested locations, the progress evident in addressing timber harvests, a topic with a highly divisive history, suggests the potential for progress on some of these additional contested issues in the future.

Conclusion

Natural resource management seeks to achieve multiple objectives. While in some cases these objectives may be complementary, in many cases they involve trade-offs that are contested by different stakeholders. The emphasis of landscape-scale, forest restoration efforts on large geographic areas, and with outcomes that will likely span a long-time horizon, increased the ecological and social complexity of these decisions. However, at the same time, this long-term, large-scale focus also appeared to encourage participants to think more broadly about their desired outcomes on the landscape. By thinking about restoration at this larger level, participants were able to move beyond the more typical conflicts associated with specific resource management decisions, and more easily recognize areas of common ground in their visions of future conditions, such as improved resilience.

Moreover, findings here suggested that reaching agreement on desired future conditions allowed participants to discuss specific management actions and, in some cases, support actions they likely would have viewed as incompatible with their preferences prior to engaging in the CFLRP efforts. While there is still considerable work to be done, ecological restoration can provide a common goal on which many stakeholders, even those with typically disparate perspectives like industry and environmental organizations, can agree. Moreover, the relationship outcomes achieved through the collaborative efforts suggests the CFRLP approach has the potential to provide a venue to work through topics where disagreement may still remain.

That said, this analysis also identifies a current gap in the collective conceptualization of restoration within our study locations. Despite the emphasis on social, economic, and ecological outcomes in the guiding legislation that established the CFLRP, our analysis found that ecological aspects of restoration are emphasized to a greater extent than associated social or economic considerations. This finding suggests that additional attention is needed to ensure success in achieving the full range of stated CFLRP objectives.

References

Ackermann, F., Cropper, S., and Eden, C. (1992) 'Getting Started with cognitive mapping', *7th Young OR Conference Tutorial Papers*. OR Society, Birmingham

Altman, D. (1991) *Practical Statistics for Medical Research*. London: Chapman and Hall

Antuma, J., Esch, B., Hall, B., Munn, E., and Sturges, F. (2014) *Restoring forests and communities: lessons from the Collaborative Forest Landscape Restoration Program*. The National Forest Foundation, Montana, USA

Barro, S.C. and Bright, A.D. (1998) 'Public views on ecological restoration: A snapshot from the Chicago area', *Ecological Restoration*, 16(1): 59–65

Byrch, C., Kearins, K., Milne, M., and Morgan, R. (2007) 'Sustainable "what"? A cognitive approach to understanding sustainable development', *Qualitative Research in Accounting & Management*, 4: 26–52.

DuPraw, M.E. (2014) 'Illuminating capacity-building strategies for landscape-scale collaborative forest management through constructivist grounded theory'. PhD Thesis Dissertation, Nova Southeastern University at Fort Lauderdale, USA

Egan, D., Hjerpe, E.E., and Abrams, J. (2011) *Human dimensions of ecological restoration*. Washington DC: Island Press

Gobster, P.H. and Hull, R.B. (2000) *Restoring nature: perspectives from the social sciences and humanities*. Washington DC: Island Press

Gray, B. (1989) *Collaborating: Finding common ground for multiparty problems*. San Francisco: Jossey-Bass

Jones, N., Ross, H., Lynam, T., Perez, P., and Leitch, A. (2011) 'Mental models: an interdisciplinary synthesis of theory and methods', *Ecology and Society*, 16(1), article 46

Keele, D.M., Malmsheimer, R.W., Floyd, D.W. and Perez, J.E. (2006) 'Forest Service land management litigation 1989–2002'. *Journal of Forestry*, 104(4): 196–202

Margerum, R.D. (2011) *Beyond consensus: Improving collaborative planning and management*. Cambridge, MA: MIT Press

Martin, D.M. (2017) 'Ecological restoration should be redefined for the twenty-first century', *Restoration Ecology*, 25(5): 668–673

Moote, M.A. and Becker, D. (2003) *Exploring barriers to collaborative forestry: report from a workshop held at Hart Prairie, Flagstaff, Arizona, September 17–19, 2003*. The Ecological Restoration Institute, Flagstaff, AZ

Moote, A. and Lowe, K. (2008) *What to expect from collaboration in natural resource management: A research synthesis for practitioners*. The Ecological Restoration Institute, Flagstaff, AZ

Morgan, M., Fischhoff, B., Bostrom, A., and Atman, C. (Eds.) (2002) *Risk communication: A mental models approach*. New York: Cambridge University Press

Rubin, H.J. and Rubin, I.S. (2005). *Qualitative interviewing: The art of hearing data*. Thousand Oaks, CA: SAGE Publications

Society for Ecological Restoration Science and Policy Working Group. 2002. *The SER primer on ecological restoration*. Society for Ecological Restoration, Washington DC

Society for Ecological Restoration. (2004) *The SER International Primer on Ecological Restoration, version 2*. Society for Ecological Restoration, Washington D.C.

Thompson, J.R., Duncan, S.L., and Johnson, K.N. (2009) 'Is there potential for the historical range of variability to guide conservation given the social range of variability', *Ecology and* Society, 20(1), article 18

Toman, E., Curtis, A., Brenkert-Smith, H., Rogers, M., and Stidham, M. (2015) *Managing multi-functional landscapes at the interface of public forests and private land: advancing understanding through a comparison of experience in the US and Australia*. Final Project Report, Project ID 12–2-01–59. Retrieved from: www.firescience.gov/projects/12-2-01-59/project/12-2-01-59_final_report.pdf

Vining, J., Tyler, E., and Kweon, B.S. (2000) 'Public values, opinions, and emotions in restoration controversies' Pages 143–162. In Gobster, P.H., and Hull, R.B. (Eds.) *Restoring nature: Perspectives from the social sciences and humanities*. Washington DC: Island Press

Walpole, E.H., Toman, E., Wilson, R.S., and Stidham, M. (2017) 'Shared visions, future challenges: a case study of three Collaborative Forest Landscape Restoration Program locations', *Environment and Society*, 22(2), article 35

Wondolleck, J.M., and Yaffee, S.L. (2000) *Making collaboration work: Lessons from innovation in natural resource management*. Washington DC: Island Press

7 Challenges and opportunities for collaborative adaptive management in forest landscape restoration

Antony S. Cheng, Gregory H. Aplet, and Amy E.M. Waltz

Introduction

Implementing large-scale forest restoration programs on multiple-use lands managed by the US Forest Service is fraught with complexity, uncertainty, and legal, political, and social conflict (Butler *et al.*, 2015). A widely advocated pathway to overcoming these challenges is to employ an adaptive management approach involving individuals and organizations representing a diversity of interests, knowledge, and assets – often termed 'collaborative adaptive management' (CAM). In principle, CAM engages all interested parties in a continuous cycle of setting goals, identifying assumptions and uncertainties, implementing a range of management actions, monitoring the effects of actions, and evaluating the need for changes of goals and management actions based on monitoring results. Situating this learning-by-doing cycle within a collaborative process is theorized to spur innovation, stakeholder accountability, trust-building, and more rapid, pro-active changes to management as new knowledge is generated and integrated into decision-making (Sabine *et al.*, 2004; Armitage, 2005; Scarlett, 2013).

Although CAM in national forest ecosystem management has been advocated since the early 1990s (Bliss *et al.*, 2001; DellaSala *et al.*, 2003; DeLuca *et al.*, 2010; National Committee for Implementation of the National Forest System Land Management Planning Rule, 2018), policies have not compelled managers to sustainably fund and implement CAM programs (Biber, 2011; Schultz and Nie, 2012). This changed with the establishment of the Collaborative Forest Landscape Restoration Program (CFLRP) in 2010. The CFLRP directed Forest Service appropriated funds to implement collaboratively developed, science-informed forest restoration plans on National Forest System lands. Moreover, the CFLRP unambiguously required collaborative monitoring and adaptive management throughout the ten-year life of CFLRP projects and for five years beyond project completion (Schultz *et al.*, 2012). A study of CFLRP project monitoring approaches found that projects dedicated about 10 per cent of annual project funds to monitoring (Schultz *et al.*, 2014).

Given this unprecedented policy, program, and investment to institute CAM, it is timely to ask the following questions:

- To what extent has CAM been operationalized into practice?
- What factors facilitate or frustrate CAM performance?
- What strategies can reduce frustrations and promote progress within and beyond CFLRP?

In this chapter, we address these questions by drawing on our 'practice-based knowledge' (Weber *et al.*, 2014) stemming from our respective involvement in three CFLRP projects: the Colorado Front Range, Four Forest Restoration Initiative (4FRI), and the Uncompahgre Plateau. After describing CAM functioning for each project, we examine factors that affected the extent to which CAM is practiced across all three projects, and identify strategies for sustaining enabling factors and for reducing the impact of inhibiting factors. We end by reflecting on the implications of CAM in CFLRP projects and for CAM in multiple-use, national forest management in general.

Collaborative adaptive management in practice

Colorado Front Range

The collaborative monitoring plan for the Colorado Front Range CFLRP was completed in June 2011 by a group composed mostly of national forest managers and scientists from various Front Range forestry and fire-related organizations (Clement and Brown, 2011). The plan specified desired trends in forest conditions in qualitative terms and the methods that would be used to determine if treatments were moving the forest in the desired direction. The drafting process proved more difficult than expected, as the team quickly realized they were jumping into the middle of a process without having established precise goal statements, desired conditions, or even a detailed description of the nature of treatments, all of which would have made it much easier to write a monitoring plan. In addition, trust between stakeholders and land managers was tested early, when collaborative members raised concerns that treatments were not adhering to protections described in the law, but the group lacked a process for addressing those concerns. A primary reason for this was that the first CFLRP-funded treatments, staged for implementation in summer 2011, were based on decision documents and environmental analyses pursuant to the National Environmental Policy Act (NEPA) completed prior to the CFLRP – so-called 'shelf-stock' projects. Further, communication was hampered by the fact that many Forest Service field staff responsible for implementing these shelf-stock projects were not regular and active participants in the collaborative process.

It became clear that a more detailed description of the larger process of project planning and decision-making was needed, and a group was formed to craft a model for adaptive management. The group's report, *Collaborative Implementation of Forest Landscape Restoration in the Colorado Front Range* (Aplet *et al.*, 2014), provided a cursory introduction to adaptive management and contained a diagram detailing the steps that the monitoring team, later labelled the 'Landscape Restoration Team' (LRT), intended to use to carry out collaborative implementation of the CFLRP project. The diagram (Figure 7.1) put the monitoring plan squarely in the centre of a larger process that began with a statement of the problem, articulation of desired conditions, identification of treatment areas, description of treatment methods, and concluded with individual project planning, pre-treatment and post-treatment monitoring, and data evaluation. The purpose was to clarify the broader cycle of adaptive management, from goal-setting to feedback from data into future management actions.

During the first ten years of the Colorado Front Range CFLRP project, about 10 per cent of annual funds were allocated to monitoring. Monitoring of forest structure was conducted by Forest Service field staff, contractors, and the Colorado Forest Restoration Institute at Colorado State University (CFRI). In some cases (e.g. for wildlife), monitoring was conducted through a contract between the Forest Service and a non-profit that was external to the collaborative stakeholder group. The details of these monitoring contracts were developed outside of the collaborative group, thereby contributing to

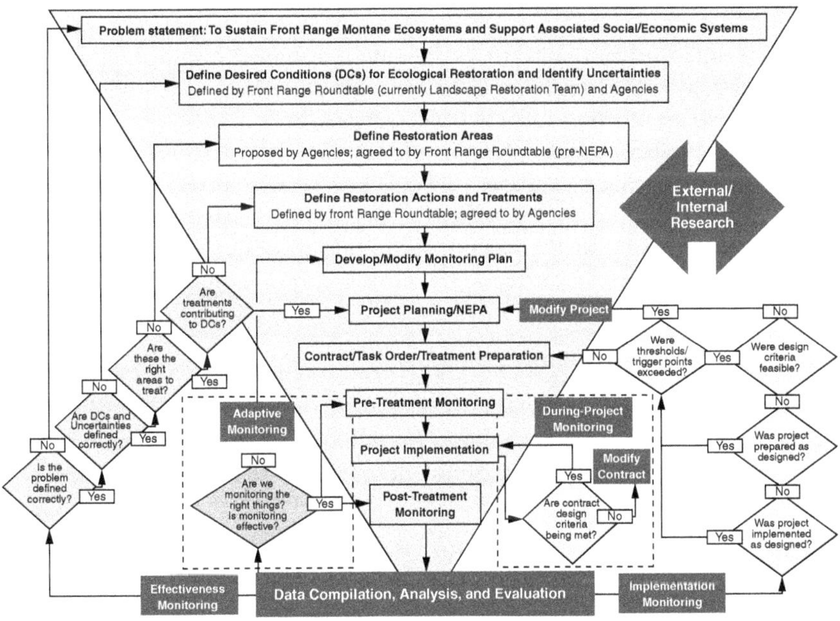

Figure 7.1 The Colorado Front Range CFLRP adaptive management model.

Source: Aplet *et al.*, 2014 (used with permission).

the cloud of budgetary and administrative opacity that hung over the project. The LRT, which also conducted field trips to treatment sites, met annually to evaluate the data using an evolving set of protocols, with the intent of making recommendations back to the Forest Service.

In practice, inadequate data and a number of social and institutional barriers limited progress. Because an existing ten-year stewardship contract and the aforementioned shelf-stock were already in place, new project planning was not needed to begin implementation, but modification of treatments was restricted to those changes that could be instituted within the limits of plans that predated the CFLRP. Still, much progress was made through the informal translation of collaborative recommendations into individual 'task orders' of the stewardship contract. In 2012, a sub-team of the Roundtable initiated planning for a new 27,500 hectare (ha) (68,000 acres) landscape within the CFLRP project area that created an opportunity for fresh project design under a new NEPA process. As input into the project development and environmental analysis process, the collaborative group prepared a set of recommended actions and recommendations for adaptive management and monitoring. The project decision, finalized in July 2017, presented an additional opportunity to enact CAM practices, although at the time of writing project implementation had not yet begun.

Staff turnover and retirements, especially among Forest Service participants, eroded group cohesion, trust, and shared understandings of collaborative goals and practices. In turn, collaborative progress towards defining more meaningful desired conditions stalled and required the retracing of steps. Even when the monitoring results were presented at annual monitoring reviews, those meetings ended with no clear strategy for what, why, and how the results of this monitoring, and any other new research on forest restoration by academic or federal researchers, would be incorporated into subsequent project implementation decisions. Several key participants (e.g. line officers and field implementers) new to the CAM process – either because of their lack of understanding or their personal attitude towards the CAM process – exhibited little interest in clarifying how this incorporation might occur.

Four Forest Restoration Initiative (Arizona)

In the six years following the 4FRI's selection as a CFLRP project, the 4FRI Stakeholder Group and other entities outside of the group worked with the Forest Service on the first Environmental Impact Statement (EIS) (USDA Forest Service, 2016) identifying over 240,000 ha (600,000 acres) for forest restoration treatments. During this time period, a stakeholder working group (eventually named the Landscape Assessment and Monitoring Working Group – LAM) worked through multiple versions of monitoring and adaptive management plans with the Forest Service. Within this smaller working group, a main concern was the extent to which Forest Service implementation would achieve the desired ecological conditions. The 4FRI working group was sceptical that the Forest Service's use of traditional silvicultural techniques would successfully achieve collaboratively developed ecological desired conditions at both stand (40–100 ha) and the landscape (4,000–10,000 ha) levels. In the face of annual reporting requirements, short-term metrics were also needed to better communicate progress to local, state, and national elected officials and audiences.

The monitoring and adaptive management plan (AM plan) was developed to be included in the first EIS specifically to: (1) monitor project implementation; (2) monitor progress towards collaborative desired conditions; (3) assess effectiveness at meeting the intent of the Forest Landscape Restoration Act; and (4) provide thresholds to trigger adaptive management considerations. The emphasis on the adaptive management plan reinforced the need for quantifiable objectives and identifiable triggers or thresholds where adaptive management action would be initiated in revised treatments.

To assure that adequate metrics were used to assess trends, indicators were selected based on attributes that were easily measured, precise, sensitive to changes over time, and satisfied multiple objectives of the monitoring process. Once the indicators were selected, triggers (sometimes described as thresholds) were identified that would signify movement towards an undesired outcome; triggers were meant to help indicate whether or not a change in management was advisable. When there was sufficient information to develop a threshold that suggested a trend away from the desired conditions, the AM plan went on to describe and outline the potential adaptive management actions (Figure 7.2). When a trigger or threshold was reached, the monitoring framework focused on the need to assess if or how management actions had contributed to the outcomes. Results from these questions would potentially result in an adjustment to the EIS (see Figure 7.2, 'Change Needed' box).

The AM plan also included the designation of a 'Stakeholder Multi-Party Monitoring Board' to oversee and implement monitoring and ensure data analyses and summaries were done to assess any need for adaptive management. The board's duties included developing adaptive management recommendations (with stakeholder group approval) for final presentation to the Forest Service. The intent was that the Forest Service and the multi-party monitoring board collaboratively evaluate the monitoring data and other relevant data to establish causal relationships. Based on the evaluation, follow-up actions would be developed. These could include, for example, continued monitoring, collecting more refined data, implementing the existing adaptive management action, or developing a new adaptive management action.

Due to a lack of trust, both among stakeholders and between stakeholders and the Forest Service, the 4FRI Stakeholder Group's activities were spelled out in explicit detail. The 4FRI adaptive management process identified key participation roles for the Stakeholder Group, which could choose to recommend adaptive management actions to the Forest Service. In addition, the process was formally institutionalized by its inclusion in the project's final EIS and Record of Decision. However, the Forest Service implementation timeline

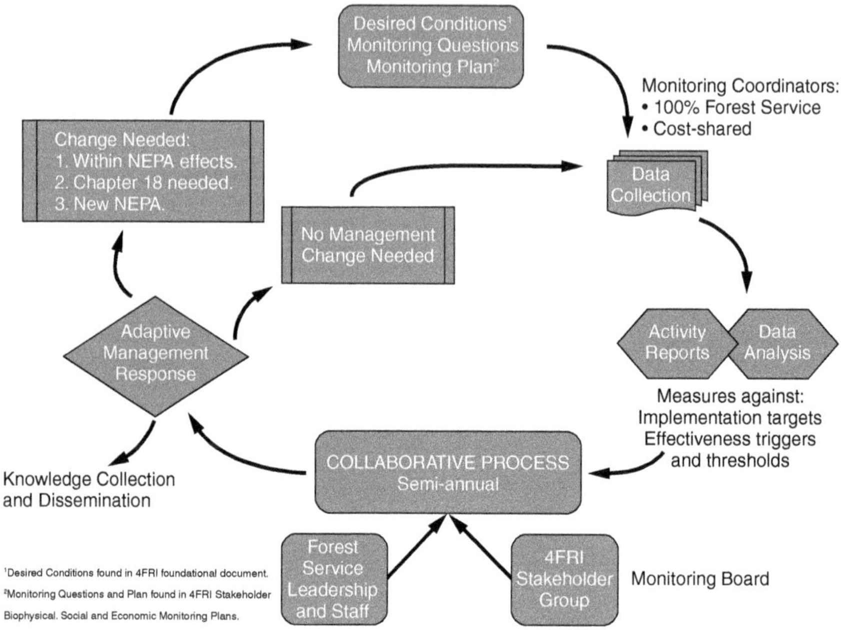

Figure 7.2 The Four Forest Restoration Initiative CFLRP adaptive management model.

Source: USDA Forest Service, 2016 (used with permission).

created a gap between the signed EIS in 2015 and the first treatments that could be monitored to inform adaptive management in 2018.

As of 2018, 4FRI forest restoration treatments had not been accomplished on a scale sufficient for the AM plan to be fully enacted beyond implementation metrics. Significant time-lags between the final signed Record of the Decision for an EIS and subsequent implementation and monitoring are not uncommon in federal land management. However, this delay, combined with both 4FRI stakeholder and Forest Service turnover, resulted in the lack of a consistent 'champion' for 4FRI adaptive management. More veteran stakeholders continued to provide re-education of the AM plan for both the Stakeholder Group and the Forest Service. But with so few treatments completed, it is too early to say if the inclusion of adaptive management in a signed Record of Decision will help ensure the continued use of both the plan and adaptive management steps. Finally, adaptive management recommendations by the stakeholder group could not be guaranteed to be incorporated into future work, as it was up to a Forest Service official to determine which management actions were implemented.

Uncompahgre Plateau (Colorado)

Prior to being selected in 2010 as a CFLRP project, Forest Service personnel and non-agency stakeholders had developed a multi-party monitoring strategy for a collaboratively developed forest restoration project, the 7,000 ha (17,200 acres) Uncompahgre Mesas Forest Restoration Project. This strategy included monitoring objectives, indicators, and measures; however, it did not include a well-specified strategy for how and when the collaborative learning from monitoring results would feed back into subsequent implementation decisions or new project planning efforts. After the Uncompahgre Plateau was selected as a CFLRP project, the monitoring strategy was amended and completed in May 2011 to include a description of the people responsible for monitoring, data management, and archiving plans, and a plan for communicating findings to the collaborative group, the Forest Service decision makers, and the public (CFRI, 2011). An initial list of seven monitoring topics was compiled based on a funding level of 10 per cent of total annual project funding. Monitoring of restoration treatments was conducted by a combination of Forest Service and other natural resource government agency specialists, CFRI, and interns from local high schools.

The CAM process did not involve a formal, structured group but was an open-participation and informal process. As a result, unlike the Colorado Front Range and 4FRI, the Uncompahgre did not have a formally adopted adaptive management strategy and box-and-arrow diagram. Annual monitoring meetings were conducted involving participation from a diversity of organizations and interests were convened jointly by the CFRI, the Uncompahgre Plateau Partnership, and the Forest Service. The Uncompahgre Plateau Partnership is a not-for-profit organization dedicated to ecological restoration on the plateau by pooling expertise, technologies, and funding from governmental and non-governmental organizations. Attendance over the course of the initial project averaged about thirty individuals. In addition to the annual monitoring meetings, annual field trips were conducted every summer to bring interested parties to project sites to qualitatively review restoration treatments, learn about operational constraints that affect the translation of restoration concepts to on-the-ground management activities, and visualize monitoring results at plot locations.

One of the persistent challenges facing Uncompahgre adaptive management was the turnover in participants. Another challenge was interpreting and monitoring results, and incorporating the resultant learning into subsequent decision-making in a predictable, transparent manner. On the one hand, there was general agreement that monitoring results indicated a trend towards desired conditions at the stand scale. On the other hand, a critique from a segment of the participants emerged that, at the landscape-scale, the forest restoration treatments may have had little effect, been unnecessary, and led to undesirable long-term ecological effects based on historical ecological assessments across large spatial and long temporal scales (Baker, 2017). While such critical perspectives can be essential in the CAM process to revisit questions about overarching restoration desired conditions, goals, and locations, the lack of a clearly-defined CAM process through which to deliberate and integrate new knowledge exposed a deficiency in the Uncompahgre CAM process.

Synthesis and implications

CAM elements common among landscapes

Looking across all three projects to assess the first of our three guiding questions (To what extent has CAM been operationalized into practice?), it appears that the CAM approach for the Colorado Front Range provided the highest level of detail in terms of explicit steps, activities, and guiding questions for adapting goals, locations, and design of forest restoration treatments. The 4FRI strategy and diagram were more general in terms of steps, but identified who was expected to participate and where in the process they would do so. The Uncompahgre Plateau CAM strategy outlined monitoring topics, approaches, and roles but lacked detail regarding steps, activities, and roles for feeding monitoring results back into subsequent management decision-making – relying instead on existing relationships, trust, and a history of collaboration to ensure that collaborative learning from project monitoring would be integrated into subsequent project decision-making.

We can point to specific elements, common to all three cases, which contributed to the advancement of CAM:

- The Forest Service allocated approximately 10 per cent of annual CFLRP budgets to support the development and implementation of multi-party monitoring strategies.
- CFLRP coordinators/facilitators convened annual or semi-annual meetings and field trips bringing together managers, stakeholders, and scientists to review and interpret results and recommend management changes.
- CAM participants modified monitoring strategies to include new indicators, measures, and analytical methods when participants realized monitoring results were insufficient to answer monitoring questions.

However, all three cases demonstrated a structural failure to connect monitoring results to 'plug-in points' intended to affect future management. Such disconnects were evident when stakeholder recommendations based on monitoring failed to influence:

- Restoration prescriptions for phased projects.
- Plans for subsequent projects.
- Project goals, desired conditions, and treatment locations.
- Modifications to the monitoring plan.

In short, while each CFLRP developed well-defined CAM strategies, actual practice fell short of expectations due to the lack of formal connections where collaborative learning from monitoring and research should have been inserted into Forest Service decision-making. The extent to which progress was made depended on the willingness of *individuals within the Forest Service* to step out of their normal organizational roles to meet the intent of CAM. The importance of key individuals choosing to try new approaches is not surprising and is well-documented in organizational change and innovation research (Burke, 2002; Steelman, 2010). However, this reliance on individual choice may also indicate that CAM across our CFLRP cases, despite having dedicated financial resources and seven years of implementation, was not institutionalized and was prone to variations in performance.

Factors that facilitate or frustrate CAM performance

This variation in CAM performance across our three CFLRP cases invokes our second guiding question: What factors facilitate or frustrate CAM performance? On the facilitating side, having a policy mandate requiring multiparty monitoring compelled the Forest Service to allocate funding to support monitoring activities. That this funding was flexible and sustained over many years was key to addressing time-lags that occur between a final project decision and actual implementation. In turn, in each case these funds attracted a core group of interested, dedicated individuals who valued collaborative monitoring and adaptive management, including the sustained participation of subject matter specialists from various research institutions and management agencies. The funding also facilitated the engagement of boundary-spanning, bridging organizations that were essential to initiating and sustaining the CAM processes over time.

Factors frustrating CAM practices in our cases are more nuanced and complex. We organized our diagnosis into three categories: organizational factors, scale incompatibilities, and data incompatibilities and stewardship issues.

Organizational factors

Organizational factors refer to the values, norms, rules and procedures, resources, routines, and identities expressed through the behaviours of individuals working within an organizational environment (Weick, 1995). CAM can be characterized as an emergent organizational environment in which individuals are attempting to institute a new set of activities around common values and objectives, all while still being tethered to their home organizations (Agranoff, 2006; Kettl, 2006). As a result, efforts to implement CAM can expose tensions in the participants' expectations for what CAM actually entails (Butler, 2013; Davis *et al.*, 2017). We identified three categories of organizational factors among our cases that contributed to common tensions:

1 Weak commitment to collaboration: CAM participants varied in their goals for, and subsequent commitment and accountability to, CAM. For example, we witnessed agency managers and stakeholders who participated in the CFLRP primarily as a means to secure funding to implement shelf-stock projects. For these participants, the CFLRP funding accelerated the production of 'acres treated' – a valued annual performance metric for Forest Service managers (Smith *et al.*, 2011). There appeared to be an ambivalence about, and even an aversion to, critically questioning the science basis underpinning restoration goals and treatment designs. Additionally, in several instances stakeholders advanced recommendations based on monitoring results regarding proposed actions or restoration silvicultural prescriptions, expecting that the recommendations would be incorporated into subsequent decisions. When the stakeholders later discovered that this input was not incorporated into the final decision, managers explained that the recommendations would not be utilized because the agency would not modify existing projects, or perceived that the agency was prohibited from doing so by the Federal Advisory Committee Act. This led to ongoing, unresolved disputes over what constitutes 'collaboration' in the CFLRP context.

2 A gap between collaborators and implementers: The functional separation between those agency staff involved in collaborative processes and those implementing projects on the ground contributed to the lack of 'plug-in points' for collaborative feedback. In the 4FRI and Uncompahgre cases, implementers – those Forest Service staff responsible for designing, preparing, and administering forest restoration operations – were not involved in working collaboratively with stakeholders during the NEPA process. In the Colorado Front Range, the NEPA analyses and decisions for the portfolio of projects comprising the CFLRP were completed several years prior to 2010, without collaborative engagement. Projects reflected collaborative input only to the extent that high-level planners and silviculturists had written broad collaboratively-developed goal statements into prescriptions; implementers were often unaware of a collaborative vision for restoration. When projects failed to meet stakeholder expectations and recommendations were made to modify treatments, field staff and contractors/operators were surprised to learn that stakeholders expected reciprocal collaborative behaviours. Moreover, adaptive management guidelines in signed environment documents (e.g. 4FRI 1st EIS) may not be implemented, particularly given that the Forest Service commitments to monitoring and adaptive management plans in an EIS are not legally enforceable (Schultz and Nie, 2012).

3 The third organizational factor is turnover in CAM participants. The Forest Service has a long-standing practice of moving personnel frequently, but an obvious impact was that new personnel were unaware of the collaborative groups' goals, processes, and behavioural expectations for CAM. The loss of participants with vital technical skills also

contributed to delays or periods where no feedback of monitoring results occurred. Perhaps more profound is that new participants may feel little or have no imperative or incentive to honour previous collaborative understandings and agreements, effectively derailing the process. In one case, a new Forest Service line officer exhibited this behaviour, putting in jeopardy the CAM process to which the officer's predecessor had committed.

Scale incompatibilities

Scale incompatibilities are caused by mismatches in spatial and temporal scales between monitoring results and decision-making, as well as the different organizational levels at which forest restoration decisions are made. We identified three categories of scale incompatibilities that affected CAM performance across our cases:

1. A mismatch between the spatial scales of monitoring and the spatial scales of restoration goals. Most forest restoration monitoring indicators and data collection across the three CFLRP cases were at the forest stand scale, an area of between ten and forty hectares. Monitoring measures and results on this scale answers questions about changes in fine-scale patterns of forest vegetation characteristics. However, many of the questions of greatest interest across the CFLRP cases are relevant at the landscape-scale (ranging from thousands to hundreds of thousands of hectares), such as fire spread, wildlife populations, and watershed effects. It was not until 2017, seven years into the Colorado Front Range and Uncompahgre Plateau projects, that analytical methods and tools were adapted to utilize aerial imagery for answering landscape-scale questions. By then, the projects were nearly over, leaving little time for the CAM process to complete the feedback loop and revisit landscape-scale restoration goals, assumptions, and treatment designs.
2. Different timeframes for project implementation, monitoring results, and planning new restoration projects. This incompatibility made it challenging to provide input based on monitoring results into subsequent project decisions about goals, desired conditions, and locations and types of treatments. By the time the CAM participants reviewed monitoring results, managers had already changed subsequent project designs and prescriptions, based on meetings, field trip discussions, and informal conversations with individual collaborative members. Generally speaking, the delay between project implementation and the evaluation of monitoring results is incompatible with the timely implementation of new projects, which are driven by annual budget imperatives and performance targets.
3. Differing participants across organizational levels: In the cases of CFLRP, projects involving multiple national forests, such as the Colorado Front Range and the 4FRI, district-level line officers, specialists,

and implementers were often left out of the CAM loop because national forest-level personnel acted as Forest Service representatives in the CAM processes. In such cases, on-the-ground implementers and district-level staff were often unaware that they had any roles in the CAM process and feedback loop, implying a need for better communication and coordination across organization levels when dealing with landscape restoration.

Data incompatibilities and stewardship issues

The monitoring data across our three CFLRP cases was collected by multiple entities, each with their own data management systems that were not necessarily compatible with one another. For example, forest stand-level data collected by the Forest Service typically are input into a corporate database called 'FSVeg', but non-Forest Service participants in CFLRP monitoring did not have ready access to FSVeg and maintained their forest stand-level data in their own databases. Spatial data and analysis results, such as those derived from aerial imagery and remote sensing, were similarly problematic. Further, institutional computer network servers on which the monitoring data was stored had protective firewalls that prevented non-institutional members from accessing and contributing to the data; these data-sharing efforts were necessary to sustain CAM both over time and across monitoring measures.

While none of these issues would be impossible to overcome, the barriers were sufficiently high to inhibit cooperative and collective action by the parties engaged in monitoring. Establishing a new collaborative governance arrangement to address CFLRP monitoring data incompatibility and stewardship issues exceeds the capacity of the CFLRP participants. However, this issue will need some resolution for future collaborative monitoring efforts.

Policy and practice implications

From our vantage point as stakeholders embedded in three CFLRP projects over many years, there is reason to celebrate progress in CAM across our three CFLRP cases. We critically assessed those factors that limited CAM (which are sometimes institutionalized) and expected behaviour in forest landscape restoration and stewardship on National Forest System lands. For future collaborative monitoring efforts, we propose the use of four specific remedies, which were built off ideas advanced by others engaging in collaborative national forest governance and recent innovations in attempts to systematize CAM beyond individual choices (our recommendations echo those of the National Committee for Implementation of the National Forest System Land Management Planning Rule (2018)).

Recommendation 1: Develop a 'chartering' document that provides detail on the timing and sequence of actions, and identify those who are responsible for those actions for each step of the CAM process

While box-and-arrow diagrams and general conceptual descriptions are starting points for defining CAM, additional detail is needed about the timing, sequence, and roles pertaining to specific actions in each box and arrow. Such details can be located in a document that serves as a type of CAM process charter and can function as a central organizing principle for stakeholders, especially in the face of personnel turnover. These chartering documents are more likely to be effective through the inclusion of 'will' and 'shall' statements incorporated into collaboratively developed and agreed-upon CAM strategies and diagrams – formally adopted within final Forest Service project decision documents to communicate expectations of CAM actions and behaviours on the part of both the Forest Service and the public. For interested readers seeking an example, the Record of Decision for the Upper Monument Creek EIS (USDA Forest Service, 2017) includes this type of 'will/shall' language.

Recommendation 2: To address further CAM process details and the participant turnover issue – create and sustain a predictable process for reiterating all parties' commitments to CAM processes

A common approach to institutionalizing expected behaviours of individuals within collaborative arrangements is to develop memoranda of agreement or understanding (MOA or MOU, respectively) that are then signed by participating organizations. Similar to integrating 'shall' language in a Record of Decision or Decision Notice, the benefit of a signed MOA/MOU is that it documents the commitments made by all parties. While a MOA/MOU may not be legally binding, it is nevertheless a document that can endure through personnel changes and can be a mechanism by which existing participants can convey to new participants – both Forest Service and stakeholders – CAM expectations. Such a document was signed by participants of the stakeholder group and supervisors of each national forest involved in the 4FRI (USDA Forest Service, 2011). Since the 4FRI had yet to generate projects sufficient to produce monitoring results, the MOU remains untested at the time of this writing. However, the presence of MOA/MOU documents provided an opportunity for all participants to annually renew commitments to the CAM process.

A MOA/MOU document can also function as a type of operational 'business plan' document specifying CAM process details left out of a Record of Decision or Decision Notice. In addition, and to address the impacts of personnel turnover on collaborative processes, the Forest Service issued guidance in 2006 for developing a 'handover memo' for departing personnel (especially line officers) to notify incoming personnel about notable commitments made to collaborative groups and partnerships (National Forest Foundation, 2008). Collaborative stakeholders and Forest Service personnel could use the memo to communicate their expectations to incoming personnel. Similarly, stakeholders could adopt a parallel process for addressing their own turnover challenges so as to reciprocate Forest Service efforts.

Recommendation 3: Ensure that agency leadership sets the tone and supports innovation and a license to learn

Forest Service line officers can set the tone for collaborative learning by simply being present and engaged when collaborative learning from monitoring and research occurs. Such engagement sets out the message, for both agency personnel and for stakeholders, that this learning and feedback are high priorities. Additionally, when adaptive management recommendations exceed the comfort level of agency staff, line officers, along with the stakeholders, can provide assurances and cover for the staff who take the risks.

To promote internal collaboration, Forest Service officials have the discretion to link budgets across resource programs (e.g. timber, wildlife, range, fuels) to instigate collaboration among resource staff – similar to how the Integrated Resource Restoration budget pilot program promoted cross-resource planning and prioritization on some national forests (Schultz *et al.*, 2015). Even though the GMUG National Forest was not under this pilot program, line officers required their resource staff to combine resource program funds to develop and implement Uncompahgre projects. In turn, this larger pool of implementation funds attracted the interest of non-agency partners to contribute funds towards the Uncompahgre project because it increased the impact of their contributions and expanded the spatial scale of restoration treatments to address their interests.

For example, Colorado Parks and Wildlife, the Mule Deer Foundation, the National Wild Turkey Federation, and the Rocky Mountain Elk Foundation contributed over $2.5 million in cash (not in-kind) towards CFLRP forest vegetation treatments between 2011 and 2017. At annual Uncompahgre CFLRP stakeholder meetings, Forest Service responsible officials, resource staff, and stakeholders were provided with the coming year's project budgets and, together, they deliberated on allocating a portion of the annual budget to monitoring. While this level of transparency and engagement may be considered a radical departure from the agency's norms, it may be a mechanism to sustain interest and tangible commitments to CAM.

An additional suggestion is for the Forest Service to adopt the norm of external feedback for periodic performance reviews conducted by science-based organizations, such as the Forest Service's Research & Development branch and university faculty and staff evaluations. External review may be especially important in light of research suggesting that existing performance measures may have little impact on improving the responsiveness and accountability of Forest Service personnel to policy goals like collaboration and restoration (Schultz *et al.*, 2015). If the reviews of non-agency stakeholders involved in the collaborative processes affecting a national forest unit are considered as part of that unit's performance evaluations, Forest Service personnel would then have an incentive to invest time and resources in stakeholder collaboration.

Recommendation 4: Sustain conditions in which boundary spanners can establish, evolve, and endure

Boundary spanners are individuals and organizations serving as knowledge and communication bridges across disciplinary, administrative, landownership, professional, social, and/or cultural group boundaries (Mollinga, 2010). Across our three cases, and from other CFLRP-related research, boundary spanners were central to the CAM process by convening stakeholders, contributing to monitoring data collection and analyses, and facilitating collaborative learning based on monitoring results for subsequent forest restoration design and implementation (see also Colavito's chapter in this volume).

While the roles and contributions of boundary spanners are well-known from empirical research on collaborative governance in natural resource management (Carr and Wilkinson, 2005; Breuer *et al.*, 2010), little is known about how to actually create and sustain boundary spanners to serve CAM processes over long time periods often needed to feed collaborative learning back into resource management decision-making. One potential pathway, based on our case studies, is to engage colleges and universities in every state with established outreach and engagement programs, such as Cooperative Extension or similar programs that already serve boundary-spanning functions (Breuer *et al.*, 2010). The structure, roles, and funding for these entities would need to be defined to fit regional and place-specific contexts so as to add value.

Although our experience is limited to only two states, the credibility and the sustained engagement of the Ecological Restoration Institute in the 4FRI and the Colorado Forest Restoration Institute in the two Colorado CFLRPs stemmed from their institutional integration and financial support from their respective universities, which amounted to a 1 : 1 ratio to the funds the institutes receive through the Forest Service. Several other CFLRPs engage similar university-based programs to serve boundary-spanning roles in their CAM processes. There is potential to build on the CFLRP experiences with university-based entities serving as boundary spanners in CAM processes to address Forest Service program needs, such as the provisions of the 2012 national forest planning rule calling for collaboration, monitoring, and adaptive management for land and resource management plans.

Conclusion

The CFLRP was a policy experiment to address ecological, social, and economic challenges associated with multi-use national forest lands in the United

States; this policy was the first to create the conditions in which collaborative adaptive management (CAM) could be applied in practice. To this end, this research on the CFLRP offers important lessons on how policy-makers, responsible agency officials, and the community of stakeholders interested in, and affected by, management decisions can create durable, institutionalized change in adopting and diffusing CAM. CAM is not just important to collaborative restoration projects but, as we noted above, it is also being embraced more broadly in the 2012 planning regulations under the National Forest Management Act, which directs the Forest Service to integrate collaboration, assessment, monitoring, and adaptive management as the overarching framework for revising long-term land and resource management plans (36 CFR §219 et seq.). As the agency embraces CAM throughout its efforts, there is a role for practitioners and researchers to continually engage in critical self-analysis and collaborative learning to inform practice-based knowledge, policy learning, and scholarship on CAM and its supporting conditions.

References

Agranoff, R. (2006) 'Inside collaborative networks: ten lessons for public managers', *Public Administration Review*, 66 (Supplement): 56–65

Aplet, G., Brown. P, Briggs, J., Mayben, S., Edwards, D., and Cheng, T. (2014) 'Collaborative implementation of forest landscape restoration in the Colorado Front Range', Colorado Forest Restoration Institute, Colorado State University, Fort Collins, CO. https://cfri.colostate.edu/wp-content/uploads/2016/05/2014_Aplet-et-al.-FR-CFLRPAdaptiveManagement_CFRI-TB-1403.pdf

Armitage, D. (2005) 'Adaptive capacity and community-based natural resource management', *Environmental Management*, 35(6): 703–715

Baker, W.L. 2017. The landscapes they are a'changing: Severe 19th century fires, spatial complexity, and natural recovery in historical landscapes on the Uncompahgre Plateau. Fort Collins, CO: Colorado Forest Restoration Institute, Colorado State University. https://cfri.colostate.edu/wp-content/uploads/sites/22/2017/10/2017-03-Baker-Uncompahgre-GLO-report_FINAL_CFRI_1704.pdf

Biber, E. (2011) 'The problem of environmental monitoring', *University of Colorado Law Review*, 83(1): 1–82

Bliss, J.G., Aplet, G., Hartzell, P., Harwood, P., Jahnige, D., Kittredge, S., Lewandowski, and Soscia, M.L. (2001) 'Community-based ecosystem monitoring', *Journal of Sustainable Forestry*, 12(3–4): 143–167

Breuer, N.E., Fraisse, C.W., and Cabrera, V.E. (2010) 'The Cooperative Extension Service as a boundary organization for diffusion of climate forecasts: a 5-year study', *Journal of Extension*, 48(4): 4–7

Burke, W.W. (2002) *Organizational change: theory and practice*. Thousand Oaks, CA: SAGE Publications

Butler, W.H. (2013) 'Collaboration at arm's length: navigating agency engagement in landscape-scale ecological restoration collaboratives', *Journal of Forestry*, 111(6): 395–403

Butler, W.H., Monroe, A., and McCaffrey, S. (2015) 'Collaborative implementation for ecological restoration on US public lands: implications for legal context, accountability, and adaptive management', *Environmental Management*, 55(3): 564–577

Carr, A. and Wilkinson, R. (2005) 'Beyond participation: boundary organizations as a new space for farmers and scientists to interact', *Society & Natural Resources*, 18(3): 255–265

Colorado Forest Restoration Institute (CFRI) (2011) 'Multi-party monitoring for the Uncompahgre Plateau Collaborative Forest Landscape Restoration Project – May 2011'. Colorado Forest Restoration Institute, Colorado State University, Fort Collins, CO. https://cfri.colostate.edu/wp-content/uploads/2016/05/2011_UP-CFLR-MultiPartyMonitoring-ORIGINAL.pdf

Clement, J. and Brown, P. (2011) 'Front Range Roundtable Collaborative Forest Landscape Restoration Project – 2011 ecological, social, and economic monitoring plan'. Colorado Forest Restoration Institute, Colorado State University, Fort Collins, CO. https://cfri.colostate.edu/wp-content/uploads/2016/05/2011_FR-CFLRP-MultiPartyMonitoring.pdf

Davis, E.J., White, E.M., Cerveny, L.K., Seesholtz, D., Nuss, M.L., and Ulrich, D.R. (2017) 'Comparison of USDA Forest Service and stakeholder motivations and experiences in collaborative federal forest governance in the western United States', *Environmental Management*, 60: 908–921

DellaSala, D.A., Martin, A., Spivak, R., Schulke, T., Bird, B., Criley, M., van Daalan, C., Kreilick, J., Brown, R., and Aplet, G. (2003) 'A citizen's call for ecological forest restoration: forest restoration principles and criteria', *Ecological Restoration*, 21: 14–23

DeLuca, T.H., Aplet, G.H., Wilmer, B., and Burchfield, J. (2010) The unknown trajectory of forest restoration: a call for ecosystem monitoring, *Journal of Forestry*, 108(6): 288–295

Kettl, D.F. (2006) 'Managing boundaries in American administration: the collaboration imperative', *Public Administration Review*, 66 (Supplement): 10–19

Miller, C. (2001) 'Hybrid management: boundary organizations, science policy, and environmental governance in the climate regime', *Science, Technology, & Human Values*, 26(4): 478–500

Mollinga, P.P. (2010) 'Boundary work and the complexity of natural resource management', *Crop Science*, 50(2): S1–S9

National Committee for Implementation of the National Forest System Land Management Planning Rule (2018) 'Final recommendations to the Secretary of Agriculture and the Chief of the Forest Service – February 3, 2018', US Department of Agriculture, Forest Service. www.fs.usda.gov/Internet/FSE_DOCUMENTS/fseprd575909.pdf

National Forest Foundation (2008) 'Conservation connection: tool – Forest Service Handover Memo', National Forest Foundation, Missoula, MT. www.nationalforests.org/assets/pdfs/Tool-Forest-Service-Handover-Memo_NFF.pdf

Sabine, E., Schreiber, G., Bearlin, A.R., Nichol, S.J., and Todd, C.R. (2004), 'Adaptive management: a synthesis of current understanding and effective application', *Ecological Management and Restoration*, 5(3): 177–182

Scarlett, L. (2013) 'Collaborative adaptive management: challenges and opportunities', *Ecology and Society*, 18(3): 26 [online] URL: http://dx.doi.org/10.5751/ES-05762-180326

Schultz, C.A., Coehlo, D.L., and Beam, R.D. (2014) 'Design and governance of multiparty monitoring under the USDA Forest Service's Collaborative Forest Landscape Restoration Program'. *Journal of Forestry*, 112(2): 198–206

Schultz, C.A., Jedd, T., and Beam, R.D. (2012) 'The Collaborative Forest Landscape Restoration Program: a history and overview of the first projects', *Journal of Forestry*, 110(7): 381–391

Schultz, C.A., Moseley, C., and Mattor, K. (2015) 'Striking the balance between budgetary discretion and performance accountability: the case of the US Forest Service's approach to integrated restoration', *Journal of Natural Resources Policy Research*,7: 109–123

Schultz, C.A. and Nie, M. (2012) 'Decision-making triggers, adaptive management, and natural resources law and planning', *Natural Resource Journal*, 52: 443–521

Smith, N., Deal, R., Kline, J., Blahna, D., Patterson, T., Spies, T.A., and Bennett, K. (2011) 'Ecosystem services as a framework for forest stewardship: Deschutes National Forest overview', General Technical Report PNW-GTR-852. US Department of Agriculture, Forest Service, Pacific Northwest Research Station, Portland, OR

Steelman, T.A. (2010) *Implementing innovation: fostering enduring change in environmental and natural resource governance*. Washington DC: Georgetown University Press.

USDA Forest Service (2011) Memorandum of Understanding between the 4 Forest Restoration Initiative (4FRI) Collaborative Stakeholder Group Representatives and the US Forest Service Apache-Sitgreaves, Coconino, Kaibab, and Tonto National Forests. US Department of Agriculture, Forest Service, Apache-Sitgreaves, Coconino, Kaibab, and Tonto National Forests. http://4fri.org/pdfs/MOU_with_signatures.pdf

USDA Forest Service (2016) 'Final Environmental Impact Statement for the Four-Forest Restoration Initiative with errata and objection resolution modifications, Appendix E'. US Department of Agriculture, Forest Service, Coconino and Kaibab National Forests, Flagstaff, AZ. www.fs.usda.gov/Internet/FSE_DOCUMENTS/stelprd3836625.pdf

USDA Forest Service (2017) Final Record of Decision: Upper Monument Creek Landscape Restoration. In. US Department of Agriculture, Forest Service, Pikes Peak District, Pike-San Isabel National Forest/Cimarron and Comanche National Grasslands, Pueblo, CO. www.fs.usda.gov/nfs/11558/www/nepa/98359_FSPLT3_4049345.pdf.

Weber, E.P., Belsky, J.M., Lach, D., and Cheng, A.S. (2014) 'The value of practice-based knowledge', *Society and Natural Resources*, 27: 1074–1088

Weick, K.E. (1995) *Sensemaking in organizations*. Thousand Oaks, CA: SAGE Publications

8 Use of scientific information to inform decision-making on CFLRP projects

Melanie M. Colavito

Introduction

Scientific information has been used in collaborative forest restoration efforts in numerous ways to support planning and decision-making. National forest management traditionally proceeded under a paradigm of scientific management, by which scientific information and technical training served as the foundations for agency decision-making. As the goals of forest management became more complex, governance shifted to an adaptive approach, which recognizes that, while decisions about land management goals must be informed by science, there must also be room for learning, local knowledge, and stakeholder deliberation to inform management decisions in any given location (Brunner *et al.*, 2005).

The Collaborative Forest Landscape Restoration Program (CFLRP) calls for 'collaborative, science-based restoration of priority forest landscapes' as well as use of the 'best available science and scientific application tools in ecological restoration strategies' (16 U.S.C. §7303(b)(1)(C)). The law's emphasis on both collaboration and science recognizes the complex relationship between science and stakeholders, with language that attempts to strike a balance between the role of scientific information and collaborative decision-making. The CFLRP also emerged during an era that emphasized the importance of better connecting science to decision-making through the production of more actionable scientific information that meets the needs of its intended users (McNie, 2013). Despite increasing emphasis on the importance of scientific information in collaborative forest management, there is a limited understanding of how science is used in collaborative governance settings that involve federal agencies as the ultimate decision makers.

The objective of this chapter is to examine the methods by which scientific information is used in collaborative forest management, summarizing results from a study that looked at the role of science in a subset of CFLRP projects in the western United States, and to provide practice and policy recommendations for connecting science and decision-making in the CFLRP and future collaborative forest restoration efforts. Throughout this chapter, 'science' refers to the systematic process of observing and analysing natural and social processes. Scientific information is defined as the output of a rigorous scientific process.

Science in forest management: from agency origins to the era of collaboration

In US national forests there is a need to restore forest resilience in light of undesirable conditions that resulted from past management actions (e.g. fire suppression and timber harvesting), human development into forested areas, and disturbance events that are exacerbated by climate change (North *et al.*, 2015; see also the National Forest Management Act (NFMA) planning regulations at 77 FR 21162 and 36 CFR 219 et seq. [2016], also known as 'the planning rule'). Collaborative forest restoration is seen as a viable approach to improving the ecological, economic, and social sustainability of public forestlands (Schultz *et al.*, 2012). Collaborative restoration is also a 'process through which multiple stakeholders jointly explore diverse values and interests and attempt to come to some level of agreement about appropriate land management', with the goal of developing restoration approaches that are 'scientifically credible and socially acceptable' (Selig, 2012, pp. 39–40). While collaborative approaches have become relatively common in US forest management, collaborative restoration, specifically, is a relatively new approach that warrants further examination – especially with respect to relationships between management expertise and requirements, scientific information, stakeholder input, and use of science in decision-making.

Science has played a role in forest management since the 'Progressive Era', primarily as part of the technical and expert-driven ethos of scientific management (Cortner and Moote, 1999). After the Second World War there was boom in timber harvesting and increasing concern about environmental degradation. These concerns, coupled with an emphasis on transparency in decision-making with the broader public, led Congress to delineate a more explicit role for science in decision-making during the era of environmental lawmaking. The National Environmental Policy Act (NEPA) of 1969 requires federal agencies to conduct environmental impact analyses for proposed actions. The NEPA regulations explicitly call for the use of 'high quality' science (40 CFR §1500.1 [2016]). Other statutes that directly affect forest management, such as the Endangered Species Act (ESA) of 1973 and the National Forest Management Act (NFMA) regulations include language requiring the use or consideration of the 'best available' scientific information (16 USC §1533(b)(1)(A) and 36 CFR §219.3 [2016]).

Although NEPA promoted interdisciplinary analysis and transparent use of scientific information, determining what constitutes best available science and how it should be used in decision-making is challenging and often based on information availability and individual discretion (Schultz, 2008). Under the 2012 NFMA regulations, agency personnel are required to document what information was determined to be best available science, why the determination was made, and how the information was applied (USDA Forest Service, 2013). The Forest Service must also consider other sources of information, including public comments, local and tribal knowledge, agency experience, and monitoring data in forest management planning; thus, while requirements are in place to mandate consideration of best available science, the law recognizes that science is but one factor that underlies decision-making.

Stakeholder participation since the 1970s has played an increasingly important role in forest management and has changed the role of science by bringing it more into the public sphere. In particular, administrative regulations pursuant to NEPA and NFMA spurred changes in the way that actors, who are external to land management agencies, participate in land management by explicitly including a public participation process during the development of proposed agency actions (Schultz, 2008). By the 1990s, collaboration had emerged to increase stakeholder engagement in planning (Wondolleck and Yaffee, 2000). Collaboration with multi-stakeholder groups is now institutionalized in land management planning, and the planning rule contains guidance on public participation and collaboration (USDA Forest Service, 2013). Collaborative stakeholder groups govern themselves in varying ways, but one of their primary roles is to convene diverse organizations that deliberate, and then deliver broader agreement to land management agencies. Collaborative groups have access to varying sources of scientific information, but the participation of scientists in these groups varies (Colavito, 2017a). Joint fact-finding exercises in which collaborative participants work together to generate new information are often used to resolve uncertainty (Wondolleck and Yaffee, 2000). Nonetheless, collaborative groups face challenges in determining what it means to use the best available scientific information.

Developing and applying actionable scientific information

In addition to being significant for collaboration, the 1990s were characterized by efforts to better connect environmental science and decision-making, to use science to achieve meaningful outcomes, and to develop actionable scientific information to meet end user needs (Jacobs, 2002; McNie, 2013). The linear model of science in which research is delivered at the 'loading dock' of society (Cash *et al.*, 2006) was intentionally developed to isolate science from the 'value-laden world of applications' (Meadow *et al.*, 2015). However, many environmental problems cannot be solved without considering different information sources and sets of values; therefore, decision makers need access to scientific information they perceive to be credible (i.e. developed with rigorous scientific standards), salient (i.e. relevant to their decision-making landscape), and legitimate (i.e. from a source that is considered unbiased, fair, and respected) (Cash *et al.*, 2003). Connecting science to end-users requires attention throughout all stages of knowledge management, from the design of scientific inquiries to communication and application in a management context. Understanding how to effectively develop and apply scientific information for decision-making is especially important in the context of collaborative forest restoration where conflict may exist, which can impact how scientific information is interpreted or where scientific information may be limited for certain resources or scales of interest (Larson *et al.*, 2013; Urgenson *et al.*, 2017).

To elevate scientific information in environmental decision-making and make research results and implications more actionable, significant effort has been devoted to improving science delivery, which is the communication and dissemination of scientific information to diverse audiences (Dabelko, 2005; Toman *et al.*, 2006). In natural resource management, conventional modes of science delivery, such as peer-reviewed papers, may be difficult to access and interpret for natural resource managers and collaborative groups alike (McKinley *et al.*, 2012). The use of multiple, written and oral science-delivery formats is often necessary to reach a broad audience and enhance information credibility, salience, and legitimacy (Colavito, 2017b). In politicized environments, limited attention to science and value judgments about the credibility of scientific information sources can affect what stakeholders believe (Lupia, 2013). Therefore, effective science delivery also requires attention to the format and context in which scientific information is developed and applied, which can affect its utility and relevance for end users (McNie, 2007).

Dilling and Lemos (2011) describe the ideal development and application of science as a pull-push process where the push side (i.e. science development and delivery) and the pull side (i.e. science use) must 'bridge any differences that might exist between what scientists might think is useful and what is actually usable in practice.' There must be a sustained, two-way communication between science producers and users to ensure that the resulting information is actionable for decision-making. This iterative, two-way process has been described by some scholars as the co-production of knowledge, which, in this context, is a collaborative approach that involves science producers and users throughout research to create actionable information that is meaningful in specific contexts (Bremer and Meisch, 2017; Ferguson *et al.*, 2014; Lemos and Morehouse, 2005). Meadow *et al.* (2015) outline the principles of successful co-production as 'building ongoing relationships between scientists and stakeholders, ensuring two-way communication between groups, and maintaining a focus on the production of usable science.' In order to accomplish co-production, there must be sufficient technical and disciplinary capacity, the ability to facilitate the relationship between scientists and stakeholders, and sufficient scientific resources to complete the work in a timely manner (Meadow *et al.*, 2015). Co-production requires significant time and resources, but preliminary evaluations of co-produced projects have demonstrated its potential for positive outcomes (Wall *et al.*, 2017).

Boundary organizations – which are organizations that serve as intermediaries between different types of organizations and disciplines and operate at the interface of science, decision-making, and policy – can facilitate science delivery and knowledge co-production, though these processes can also occur through existing knowledge networks (Feldman and Ingram, 2009; White *et al.*, 2008). Boundary organizations tend to exhibit four common functions, which include: convening scientists, managers, and policymakers; translating scientific and policy knowledge; facilitating collaborations to develop actionable information and products; and mediating interactions among their participants (Cash *et al.*, 2006). Boundary spanners are the individuals who facilitate boundary organization functions. In the field of forest and fire science they have been found to be key brokers of scientific information to the management community (Hunter, 2016; Wright, 2010). Research on boundary organizations has demonstrated their ability to improve trust and their capacity to use scientific information for complex, environmental decision-making (McNie, 2013). Collaborative forest restoration – which encourages scientists, natural resource managers, and stakeholders to work together to identify the most relevant scientific information for their landscapes and determine how it might be applied – provides a useful example for better understanding the development, communication, and the application of scientific information for complex environmental decisions, and the potential role of boundary organizations.

Research questions and methods

Given the opportunities and challenges to balance science and social values in collaborative forest restoration, it is important to better understand the role of science in existing collaborative forest restoration efforts. I examined how a subset of CFLRP projects were managing the use of science to provide recommendations for integrating science in the process of doing 'scientifically credible and socially acceptable' landscape-scale forest restoration (Selig, 2012). Specifically, this chapter addresses: (1) how CFLRP projects accessed, delivered, and rated existing scientific information; (2) how CFLRP projects developed new scientific information; and (3) what factors inhibited the use of science in CFLRP projects.

I conducted research on five CFLRP projects in the western United States: the Four Forest Restoration Initiative (4FRI) in Arizona; the Zuni Mountain and Southwest Jemez Mountains CFLRP projects in New Mexico; and the Uncompahgre Plateau and Colorado Front Range CFLRP projects in Colorado. I designed my research using a modified grounded theory framework in which data collection, analysis, and synthesis were iterative with results built on prior information to allow major themes to emerge over time (Strauss and Corbin, 1990). In the first research phase, which took place from August 2013 to January 2014, I conducted twenty semi-structured interviews, observed participants, and reviewed documents. The purpose of the interviews was to learn more about the factors that affect collaboration, science, and decision-making (Dunn, 2008). I supplemented this data with participant observation at several meetings, and a review of project materials, including proposals, plans, and NEPA documents (Kearns, 2008). I coded this collective data iteratively into Nvivo so as to allow major themes about science, collaboration, and decision-making to emerge inductively (Bernard and Ryan, 2010). In the second phase of research, I developed a written survey to validate the major themes and patterns that emerged from the first phase (McGuirk and O'Neill, 2008). The survey was distributed online in October 2014 to all interview respondents, who were asked to share it with fellow project participants. I received thirty-nine survey responses that were managed in 'Qualtrics'; this platform also provided basic results and descriptive statistics. I then compared the major themes and patterns from the first and second phases of research using a cross-comparative method, in which I created a matrix consisting of major themes from the five projects to highlight similarities and differences (Miles and Huberman, 1994).

Results: science-use themes in the CFLRP

The interviews and survey provided unique data from the five CFLRP projects in the study. What follows is not a comparative analysis of the five projects – due to the small and variable samples across the projects, and the potential for overlap in interview and survey respondents. Rather, I conducted an analysis that draws on evidence from the five cases to develop general findings about access, delivery, and rating of existing scientific information, development of new scientific information, and obstacles to science use. These results are followed by a discussion and recommendations for enhancing the role of science in collaborative forest restoration.

Access, delivery, and rating of existing scientific information

Science access refers to how and from where CFLRP projects obtained existing scientific information. Scientific information was most commonly provided by university-based groups, according to 90 per cent of the survey respondents who answered a question that listed scientific-expertise providers. University-based groups refer to boundary organizations such as the Southwest Ecological Restoration Institutes (SWERI) in Arizona, New Mexico, and Colorado. Of the respondents, 85 per cent indicated that scientific information was provided by the Forest Service, and the same percentage identified other federal agencies. For example, on the Southwest Jemez CFLRP project, the Valles Caldera National Preserve provided scientific information to the CFLRP project through their monitoring program. Of survey respondents from across all projects, 79 per cent said science was provided by local scientists or experts (i.e. scientists who are independent from boundary organizations or other agencies, but who are in the same geographic area as the CFLRP project, including individual researchers and private scientists). Of the respondents, 69 per cent identified science-based non-profits (e.g., The Nature Conservancy or the Forest Stewards Guild), 67 per cent selected state agencies, 59 per cent selected the Forest Service research branch, and 54 per cent said non-local scientists or experts (i.e. scientists who were independent from boundary organizations and were not in the same geographic area as the CFLRP project). When asked to rate their satisfaction with scientific information providers, survey respondents were largely satisfied with the quality of the scientific information provided by these sources. However, respondents were most satisfied with university-based groups which interacted with the CFLRP groups as boundary organizations, followed by state agencies, other federal agencies, the Forest Service, local scientists or experts, science-based non-profits, Forest Service research, and non-local scientists or experts.

Science delivery refers to the formats with which existing scientific information was communicated in CFLRP projects. Science delivery occurred both in instances in which respondents actively sought scientific information and in instances in which respondents were passive recipients of scientific information. The most common science delivery formats identified in the survey were oral. According to 95 per cent of survey respondents, scientific information was most commonly provided through informal, ongoing, face-to-face interactions between scientists, collaborative participants, and agency representatives with expertise in forest ecology and management; 85 per cent of respondents also said that scientific information was communicated in formal presentations (e.g. one-time presentations at collaborative group meetings); and only 33 per cent of respondents indicated that scientific information was delivered via webinars, which was the only oral format not delivered in-person included on the survey. Respondents were less aligned about the use of written science communication formats. Professional reports were the most common written scientific communication format used, with 69 per cent of the survey respondents saying science had been communicated in this format. Other widely used formats included working papers (62 per cent), peer-reviewed articles (54 per cent) and fact sheets (38 per cent). However, when asked to rank preferred science communication formats in another survey question, respondents indicated that peer-reviewed articles were the most preferred, followed by informal face-to-face meetings with experts, professional reports, formal presentations, working papers, fact sheets, and webinars.

The survey results highlighted the importance of oral, in-person communication, which allows for two-way conversation and interaction. In-person communication was also a common theme in the interviews. Interview respondents noted that in-person communication was beneficial for developing common goals, describing desired future conditions, and planning and conducting monitoring, all of which were informed by scientific information. For example, one agency respondent (#1) noted, 'Sitting down at the table and discussions in the woods are good for talking about what works and what doesn't.' Field trips were also an important theme in the interviews.

Respondents explained that field trips facilitated science communication by encouraging discussion and interaction that promoted knowledge sharing and learning. One agency respondent (#2) described field trips as helpful for 'social acceptance, as people better understand how the things they've recommended actually look on the ground'.

Development of new scientific information

Another common theme in the interviews was the importance of developing new science to provide actionable information for decision-making where it did not already exist. CFLRP funding could be used for monitoring, but not research. However, this did not prohibit the development of new scientific information that was not being used for monitoring and was supported by other funding sources. When asked how they were involved in the development of scientific information, respondents provided examples of projects facilitated by independent scientists and boundary organizations to address questions raised during the CFLRP process. For example, on 4FRI, research was conducted to assess how different forest resources might respond to the implementation of diameter caps, the use of which was a compromise reached by the 4FRI stakeholder group and the Forest Service (Sanchez Meador *et al.*, 2015).

Monitoring also leads to the development of new scientific information, albeit for the specific purpose of informing implementation and adaptive management (Larson *et al.*, 2013). Respondents emphasized that monitoring, in particular, presented many scientific opportunities, as well as opportunities to inform decision-making, but it was not without challenges. Interview respondents noted that the Forest Service tended to focus on implementation monitoring (i.e. monitoring to determine whether projects were implemented as planned), while collaborative groups tended to focus on effects monitoring (i.e. monitoring for the effects of project activities), which required a higher level of scientific rigor to ensure proper experimental design. Indeed, many interview respondents noted that it was difficult to reconcile differences in the level of detail desired from monitoring. At the same time, one collaborative partner (#3) noted, '[The] desire for detailed information hasn't stopped the [Forest Service] from doing treatments, and the lack of information in some cases has encouraged people to attend field trips and do sample marking exercises.' Respondents also emphasized the difficulty of doing monitoring on a large scale and incorporating monitoring results into future decision-making in an adaptive framework – which were themes that had been identified in previous studies (Butler *et al.*, 2015; Cheng *et al.*, this volume; Larson *et al.*, 2013; Schultz *et al.*, 2014). However, devoting time and resources to monitoring is critical to using scientific information in decision-making and to achieve desired conditions (Mottek *et al.*, 2017).

Science obstacles

Science was perceived to be one of many perspectives in CFLRP projects, but incorporating science into decision-making was seen as a clear objective, according to survey respondents who answered a question about perceptions of the role of science. At the same time, there was concern among respondents that the sideboards or overarching and non-negotiable criteria created by NEPA and other legal requirements could be obstacles to the use of science. As one scientist (#4) stated, 'I see most of our science being used to say, "We are trying to put it on a trajectory toward this", but of course, they can't go there with the current regulations.' This illustrates some of the tensions that exist around navigating the use of science, policy requirements, and the role of science in collaborative processes. To that end, another theme in the interview and survey results was the lingering disconnect between science, decision-making, and implementation. As one agency respondent (#5) stated, 'There's still a disconnect between what people on the ground think and what researchers think.' In a survey question that asked respondents to indicate their level of agreement with statements about obstacles to the use of science, the top five obstacles were: (1) conflicting socio-economic or political viewpoints that inhibit the use of science; (2) it is too late in the planning process to incorporate new scientific information; (3) science is not effectively communicated; (4) decisions are limited by sideboards set in NEPA; and (5) there are conflicting scientific viewpoints and scientific disagreement. These results indicate that although there is a commitment to the use of science in CFLRP projects, there is plenty of room for clarifying the role of science, effectively delivering science, and striking a better balance between scientific information, social values, and statutory requirements.

Discussion

Striking a balance between scientific information and social values in order to produce more widely agreed-upon restoration outcomes is a foundational concept of the CFLRP and collaborative forest restoration more broadly. Each CFLRP project has unique geographic, ecological, and social characteristics, but they are all related by the broader legislation, the directing agency, and the common goal of forest restoration. There are commonalities among projects with respect to how they are accessing, delivering, and developing scientific information that can provide insight and recommendations for situating and using science more effectively in collaborative forest restoration.

The CFLRP projects referenced in this chapter primarily accessed scientific information from boundary organizations, federal agencies, and local scientists and experts (i.e. independent or private scientists); however, the respondents were most satisfied with boundary organizations as their science providers. There are a few possible explanations for this finding. First, the CFLRP projects assessed in this study were co-located with three SWERI boundary organizations in Arizona, Colorado, and New Mexico. These organizations were designed and funded to facilitate the development and translation of scientific information into actionable knowledge for forest restoration. These boundary organizations were active participants in the CFLRP due to their unique mission and, hence, they were more likely to participate than other scientists and experts who may not be incentivized to engage to the same degree. At the same time, this finding is consistent with previous research that demonstrated the importance of boundary organizations in facilitating interactions among scientists, decision-makers, and stakeholders in fire and forest management (Kocher *et al.*, 2012; Hunter, 2016). Therefore, a second explanation is that respondents were accessing and preferring boundary organizations because they successfully fulfilled their mission by convening and mediating collaborative interactions, as well as delivering and translating scientific information (Cash *et al.*, 2003; McNie, 2013; Meadow *et al.*, 2015).

The CFLRP projects assessed here are also in close geographic proximity to areas with high concentrations of natural scientists, such as the Rocky Mountain Research Station branches, the Valles Caldera Natural Preserve, the United States Geological Survey branches, and numerous universities. This may explain respondent preferences for other state and federal agencies and local scientists and experts, in addition to boundary organizations, as science providers. The presence of these scientific entities has likely contributed to the development of knowledge networks that existed prior to the creation of the CFLRP (Feldman and Ingram, 2009). It is not possible to determine whether respondent preferences for boundary organizations and other local scientists and experts would be similar were it not for their geographic proximity and organizational missions. But within the CFLRP projects assessed here, these entities played important roles in the use of science.

The ways in which science is communicated in CFLRP projects can also bring science integration into the collaborative process. Scientific information was most commonly communicated in oral, in-person formats. Such communication consisted of both formal and informal conversations between scientists, collaborative participants, and agency representatives. The common use of informal, in-person delivery of scientific information supports previous work that found that interactive science delivery formats are more effective than unidirectional formats, such as webinars (Toman *et al.*, 2006). The collaborative setting of CFLRP projects enables in-person interactions, which can facilitate understanding of how to tailor science delivery to a given audience, as well as understanding the information itself through discussion and interaction. For example, respondents felt that in-person interactions during activities like field trips facilitated science delivery – a result that is consistent with other CFLRP research (Butler *et al.*, 2015; Urgenson *et al.*, 2017). In-person interaction can also provide benefits like relationship- and trust-building over time, even in situations where there are longstanding disagreements (Coleman and Stern, 2017; Walpole *et al.*, 2017). Another possible explanation for respondent preferences for oral science delivery formats, in which recipients of the information might be active or passive depending on the context, is that informal and formal modes of interaction are representative of collaborative interactions more generally (Wondolleck and Yaffee, 2000). Collaborative groups use a variety of mechanisms to engage with stakeholders, agency representatives, and scientists, so it follows that science delivery also uses a variety of mechanisms to enhance its relevance in particular contexts (Dabelko, 2005).

While in-person science delivery was most common in this study, respondents preferred peer-reviewed articles when asked to rank science communication formats. Previous work has found that boundary spanners (i.e. individuals who facilitate the boundary organization functions of convening, translating, facilitating, and mediating) typically learn about new scientific information from peer-reviewed articles (Hunter, 2016). At the same time, other work found that natural resource managers may not be effectively accessing peer-reviewed articles (McKinley, 2012). The respondents in this study consisted of boundary spanners, natural resource managers, and other stakeholders, so the preference for peer-reviewed articles here indicates a recognition of their value in providing the best available scientific information. Peer-reviewed articles also tend to pass standards of credibility and legitimacy, and previous work has demonstrated the need for scientific information to meet these standards in order to be deemed usable in decision-making (Cash *et al.*, 2003). Furthermore, in politicized or complex environments, people tend to make different choices about what information they are willing to believe (Lupia, 2013). This further illustrates the need for credible sources of scientific information, like peer-reviewed articles, to supplement other sources of science delivery formats. A preference for peer-reviewed articles, however, does not negate the importance of oral science communication formats, but suggests that multiple formats are necessary for effective communication tailored to a given context and audience (Colavito, 2017b; Dabelko, 2005).

When new scientific information is needed, the CFLRP projects assessed here developed science through new research and monitoring. These projects were not at a stage where monitoring results were being used in an adaptive management process when the research was being conducted, though monitoring was ongoing in all of the projects to varying degrees (see Cheng *et al.*, this volume). Although respondents were not directly asked about knowledge co-production, there was evidence of it occurring in both new research development and monitoring in this study. Knowledge co-production is viewed here as collaborative research that involves science producers and users in all stages of knowledge development to produce meaningful, context-specific information (Bremer and Meisch, 2017; Meadow *et al.*, 2015; Wall *et al.*, 2017). Knowledge co-production emphasizes the production of actionable science, which requires ongoing assessment and feedback from all participants throughout the research and ongoing scientist participation (McNie,

2007; Meadow *et al.*, 2015). The results from this study demonstrate that scientists are participating in the CFLRP, but the nature of scientist participation needs to be better understood. For example, it is unclear how scientist participation varies across CFLRP projects in terms of levels of and timeframes for engagement, and how this might affect co-production processes. Boundary organizations are one of the primary mechanisms through which the co-production of new knowledge can be accomplished (Cash *et al.*, 2006; Meadow *et al.*, 2015) and their presence in this study indicates that they are already facilitating knowledge co-production by convening diverse stakeholders in an iterative process to develop, interpret, and apply actionable scientific information.

Respondents in this study indicated that incorporating science was a clear objective, but there were other social values they felt have inhibited the use of science. Previous research on collaborative forest restoration has also highlighted the challenges of integrating science and social values (Egan *et al.*, 2015; Urgenson *et al.*, 2017; Pinchot Institute, 2013). The CFLRP is fundamentally about 'collaborative, science-driven' restoration that necessitates decision-making informed by science, as well as by local knowledge and perspectives (Brunner *et al.*, 2005). However, scientific information must be sufficiently actionable while meeting the standards of credibility, salience, and legitimacy for collaborative and agency decision-making (Cash *et al.*, 2003; McNie, 2013). It takes time for scientific information to be incorporated into decision-making (Wright, 2010) and it can be difficult to measure the use of science (Hunter, 2016). Furthermore, scientific knowledge is embedded in society; it both influences and is influenced by social practices (Jasanoff, 2004). Indeed, the process of balancing differences between context-specific social values and best available science in CFLRP projects, and collaborative forest restoration more broadly, requires ongoing dialogue and learning in a sustained, iterative process (Selig, 2012). Land management agencies and collaborative groups have varying capacities and resources with respect to understanding and using scientific information, and there are inherent challenges in the process of interpreting scientific information and balancing it with other information and values (Meadow *et al.*, 2015; Wall *et al.*, 2017). However, the results from this study have illustrated that engagement with boundary organizations and the use of in-person science delivery, coupled with principles of knowledge co-production, can facilitate dialogue between stakeholders taking into account scientific information and social values to reach more widely agreeable, context-specific, collaborative forest restoration plans and outcomes.

Practice and policy implications

Lessons from this study can inform collaborative environmental management efforts where stakeholders and scientists are working in concert with land managers on efforts to balance social and ecological goals. First, boundary organizations were the most common and preferred sources of scientific information for CFLRP projects. Although the CFLRP projects assessed here are co-located with boundary organizations like the SWERIs – which are designed and funded to assist in collaborative forest restoration – boundary organizations are, more broadly, designed to facilitate interactions between scientists, decision makers, and stakeholders in the development and application of actionable scientific information. Furthermore, there are additional boundary organizations that could assist in collaborative forest restoration (e.g. the Joint Fire Science Program Fire Science Exchanges and the National Oceanic and Atmospheric Administration Regional Integrated Sciences and Assessments). Boundary organizations can help bring together diverse actors, create sustained engagement, and build trust among partners in order to navigate and interpret scientific information for complex, context-specific, decision-making. Agencies and collaborative groups should continue to work with boundary organizations to facilitate the process of balancing scientific information and social values. Furthermore, independent scientists and agency scientists not associated with boundary organizations should be more encouraged and incentivized by their own institutions to work with collaborative groups to expand the sources of scientific information available and encourage broader collaboration and knowledge co-production. The role of boundary organizations and independent scientists in collaborative processes should also be more clearly defined in order to encourage transparency and clarity in the process of developing and using scientific information.

Second, oral, in-person science delivery was the most common science communication approach used in CFLRP projects, but the respondents preferred peer-reviewed articles as the most trusted source of information. This points to a potential gap between how science is communicated and what stakeholders have access to. Therefore, science delivery should be iterative and ongoing with scientific information communicated in both oral and written formats tailored to specific contexts and audiences. Peer review articles may not be equally accessible to all stakeholders so they should be part of a delivery strategy that includes dialogue about the content and implications of science in relation to peer-reviewed research results. Boundary organizations can assist in science delivery by providing access to the best available scientific information, and to boundary spanners with specific expertise or skills in developing science delivery formats in collaboration with stakeholders and agency representatives.

Third, there is still disconnect between scientific information and social values, and it is unclear how to effectively balance different perspectives and sources of information when they are not in alignment. Collaborative approaches, such as the co-production of knowledge, can be used to produce more meaningful, context-specific, scientific information, for which collaborative groups and natural resource managers alike may have a higher level of understanding and support. Knowledge co-production can also facilitate the integration of social values into the production of new scientific information, which may contribute to greater ease in integrating science with social values. The process of knowledge development also encourages dialogue and learning among varying perspectives. At the same time, there is still value in mandating transparent consideration of the best available science so that scientific information is incorporated into decision-making and not lost among competing social values. Every collaborative effort is unique in terms of its participants and social context; there should be sufficient financial and technical support tailored to specific projects.

Conclusion

Collaborative forest restoration is inherently a learning process. Iterative and ongoing dialogue between scientists, natural resource managers, stakeholders, and policymakers will facilitate robust, place-based, decision-making that is 'scientifically credible and socially acceptable' (Selig, 2012). The CFLRP is part of an evolution in how scientific information is developed, interpreted, delivered, and applied for meaningful, action-oriented decision-making that is tailored to a given social and ecological context. Assessment of the CFLRP projects in this study provides insight into the challenges and opportunities of integrating science with other sources of information and social values. This chapter explored the ways in which CFLRP projects accessed, delivered, and rated existing scientific information; how new scientific information was developed; and what factors inhibited the use of science. The prevalence of boundary organizations in this study illustrates one way in which CFLRP projects are working to enhance science delivery and use of scientific information to inform decision-making. There are many opportunities to expand the role of boundary organizations and use the principles of knowledge co-production to facilitate the development of meaningful, context-specific scientific information that can be effectively interpreted within the unique social and ecological context of a collaborative forest restoration project. It is also important to continue to document the successes and challenges of collaborative forest restoration efforts and encourage learning throughout all phases of planning and implementation so as to understand how the integration of scientific information and social values affects ecological and social conditions on the ground.

References

Bernard, H.R. and Ryan, G.W. (2010) *Analyzing Qualitative Data: Systematic Approaches*. Thousand Oaks, CA: SAGE Publications

Bremer, S. and Meisch, S. (2017) 'Co-production in climate change research: Reviewing different perspectives', *Wiley Interdisciplinary Reviews: Climate Change*, 8(6): 22

Brunner, R.D., Steelman, T.A., Coe-Juell, L.C., Cromley, C.M., Edwards, C.M., and Tucker, D.W. (2005) *Adaptive Governance: Integrating Science, Policy, and Decision Making*. New York: Columbia University Press

Butler, W.H., Monroe, A., and McCaffrey, S. (2015) 'Collaborative implementation for ecological restoration on US public lands: Implications for legal context, accountability, and adaptive management', *Environmental Management*, 55(3): 564–577

Cash, D.W., Borck, J.C., and Patt, A.G. (2006) 'Countering the loading-dock approach to linking science and decision making: Comparative analysis of El Niño/ Southern Oscillation (ENSO) forecasting systems', *Science, Technology, & Human Values*, 31(4): 465–494

Cash, D.W., Clark, W.C., Alcock, F., Dickson, N.M., Eckley, N., Guston, Jäger, J., and Mitchell, R.B. (2003) 'Knowledge systems for sustainable development', *Proceedings of the National Academy of Sciences*, 100(14): 8086–8091

Colavito, M.M. (2017a) 'The role of science in the Collaborative Forest Landscape Restoration Program', *Journal of Forestry*, 115(1): 34–42

Colavito, M.M. (2017b) 'Utilising scientific information to support resilient forest and fire management', *International Journal of Wildland Fire*, 26(5): 375–383

Coleman, K. and Stern, M.J. (2017) 'Exploring the functions of different forms of trust in collaborative natural resource management', *Society & Natural Resources*, 31(1): 21–38

Cortner, H.J. and Moote, M.A. (1999) *The Politics of Ecosystem Management*. Washington DC: Island Press

Dabelko, G.D. (2005) 'Speaking their language: How to communicate better with policymakers and opinion shapers – and why academics should bother in the first place', *International Environmental Agreements: Politics, Law and Economics*, 5(4): 381–386

Dilling, L. and Lemos, M.C. (2011) 'Creating usable science: Opportunities and constraints for climate knowledge use and their implications for science policy', *Global Environmental Change*, 21(2): 680–689

Dunn, K. (2008) 'Interviewing', In Hay, I. (Ed.) *Qualitative Research Methods in Human Geography*. 2nd ed., pp. 79–105, Oxford: Oxford University Press

Egan, D., Stoddard, M. and Formanack, A. (2015) 'Ecological and social implications of employing diameter caps at a collaborative forest restoration project near Flagstaff, Arizona, USA', *Forest Policy and Economics*, 52: 39–45

Feldman, D.L. and Ingram, H.M. (2009) 'Making science useful to decision makers: Climate forecasts, water management, and knowledge networks', *Weather, Climate, and Society*, 1: 9–21

Ferguson, D., Rice, J., and Woodhouse, C. (2014) 'Linking environmental research and practice: Lessons from the integration of climate science and water management in the Western United States,' University of Arizona, Climate Assessment for the Southwest, Tucson, AZ, Available online at: www.climas.arizona.edu/publication/report/linking-environmental-research-and-practice; last accessed May 23, 2018

Hunter, M.E. (2016) 'Outcomes of fire research: Is science used?', *International Journal of Wildland Fire*, 25(5): 495–504

Jacobs, K. (2002) 'Connecting science, policy, and decision-making: A handbook for researchers and science agencies', NOAA Climate Program Office, Silver Spring, MD, Available online: www.climas.arizona.edu/publication/report/connecting-science-policy-and-decision-making-handbook-researchers-and-science; last accessed May 23, 2018

Jasanoff, S. (2004) 'Ordering knowledge, ordering society', In Jasanoff, S (Ed.) *States of Knowledge: The Co-Production of Science and Social Order*. (pp. 1–45). New York: Routledge

Kearns, R.A. (2008) 'Knowing seeing? Undertaking observational research', In Hay, I. (Ed.), *Qualitative Research Methods in Human Geography*. 2nd ed. (pp. 192–206). Oxford: Oxford University Press

Kocher, S.D., Toman, E., Trainor, Wright, V., Briggs, J.S., Goebel, C.O., Montblanc, E.M., Oxarart, A., Pepin, D.L., Steelman, T.A., Thode, A., and Waldrop, T.A. (2012) 'How can we span the boundaries between wildland fire science and management in the United States?', *Journal of Forestry*, 110(8): 421–428

Larson, A.J., Belote, R.T., Matthew, A., Williamson, M.A., and Aplet, G.H. (2013) 'Making monitoring count: Project design for active adaptive management', *Journal of Forestry*, 111(5): 348–356

Lemos, M.C. and Morehouse, B.J. (2005) 'The co-production of science and policy in integrated climate assessments', *Global Environmental Change*, 15(1): 57–68

Lupia, A. (2013) 'Communicating science in politicized environments', *Proceedings of the National Academy of Sciences*, 110, supplement 3: 14048–14054

McGuirk, P.M. and O'Neill, P. (2008) 'Using questionnaires in qualitative human geography', In Hay, I. *Qualitative Research Methods in Human Geography*. 2nd ed. (pp. 147–162). Oxford: Oxford University Press

McKinley, D.C., Briggs, R.D., and Bartuska, A.M. (2012) 'When peer-reviewed publications are not enough! Delivering science for natural resource management', *Forest Policy and Economics*, 21: 1–11

McNie, E.C. (2007) 'Reconciling the supply of scientific information with user demands: An analysis of the problem and review of the literature', *Environmental Science & Policy*, 10(1): 17–38

McNie, E.C. (2013) 'Delivering climate services: Organizational strategies and approaches for producing useful climate-science information', *Weather, Climate, and Society*, 5(1): 14–26

Meadow, A.M., Ferguson, D.B., Guido, Z., Horangic, A., and Owen, G. (2015) 'Moving toward the deliberate coproduction of climate science knowledge', *Weather, Climate, and Society* 7(2): 179–191

Miles, M.B. and Huberman, A.M. (1994) *Qualitative Data Analysis: An Expanded Sourcebook*, 2nd ed. Thousand Oaks, CA: Sage Publications

Mottek, L.A., Yeon-Su, K., Greco, B., Becker, D.R., Hjerpe, E.E., and Abrams, J. (2017) 'The social and economic contributions of the White Mountain Stewardship Project: Final 10-year assessment – Lessons learned and implications for future forest management initiatives', *Journal of Forestry*, 115(6): 548–558

North, M.P., Stephens, S.L., Collins, B.M., Agee, J.K., Aplet, G.H., Franklin, J.F., and Fule, P.Z. (2015) 'Reform forest fire management' *Science*, 349(6254): 1280–1281

Pinchot Institute (Pinchot Institute for Conservation). (2013) 'Independent Science Panel Report Herger-Feinstein Quincy Library Group Forest Recovery Act', Pinchot Institute for Conservation, Washington DC. Available online: www.pinchot.org/doc/476, last accessed May 23, 2018

Sánchez Meador, A.J., Waring, K.M., and Kalies, E.L. (2015) 'Implications of diameter caps on multiple forest resource responses in the context of the Four Forests Restoration Initiative: Results from the Forest Vegetation Simulator.' *Journal of Forestry*, 113(2): 219–230

Schultz, C.A. (2008) 'Responding to scientific uncertainty in US forest policy', *Environmental Science & Policy*, 11(3): 253–271

Schultz, C.A., Coelho, D.L., and Beam, R.D. (2014) 'Design and governance of multiparty monitoring under the USDA Forest Service's Collaborative Forest Landscape Restoration Program', *Journal of Forestry*, 112(2): 198–206

Schultz, C.A., Jedd, T., and Beam, R.D. (2012) 'The Collaborative Forest Landscape Restoration Program: A history and overview of the first projects', *Journal of Forestry*, 110(7): 381–391

Selig, W. (2012) 'Collaboration'. In Egan, D. and Dubay, T. (Eds.) *Breaking Barriers, Building Bridges: Collaborative Forest Landscape Restoration Handbook*, pp. 36–56. Flagstaff, AZ: Ecological Restoration Institute, Northern Arizona University

Strauss, A.L. and Corbin, J.M. (1990) *Basics of Qualitative Research: Grounded Theory Procedures and Techniques*. Newbury Park, California: Sage Publications

Toman, E., Shindler, B., and Brunson, M. (2006) 'Fire and fuel management communication strategies: Citizen evaluations of agency outreach activities', *Society & Natural Resources*, 19(4): 321–336

Urgenson, L.S., Ryan, C.M., Halpern, C.D., Bakker, J.D., Belote, R.T., Franklin, J.F., Haugo, R.D., Nelson, C.R., and Waltz, A.E.M. (2017) 'Visions of restoration in fire-adapted forest landscapes: Lessons from the Collaborative Forest Landscape Restoration Program', *Environmental Management*, 59(2): 338–353

USDA Forest Service. (2013) *Forest Service Manual, Land Management Planning Handbook, Chapter 40 Key Processes Supporting Land Management Planning*. US Forest Service National Headquarters, Washington DC

Wall, T.U., Meadow, A.M., and Horganic, A. (2017) 'Developing evaluation indicators to improve the process of coproducing usable climate science', *Weather, Climate, and Society*, 9(1): 95–107

Walpole, E.H., Toman, E., Wilson, R.S., and Stidham, M. (2017) 'Shared visions, future challenges: A case study of three Collaborative Forest Landscape Restoration Program locations', *Ecology and Society*, 22(2): 35

White, D.D., Corley, E.A., and White, M.S. (2008) 'Water managers' perceptions of the science–policy interface in Phoenix, Arizona: Implications for an emerging boundary organization', *Society and Natural Resources*, 21(3): 230–243

Wondolleck, J.M. and Yaffee, S.L. (2000) 'Making collaboration work: Lessons from innovation in natural resource management'. Washington DC: Island Press

Wright, V. (2010) 'Influences to the success of fire science delivery: Perspectives of potential fire/fuels science users', Final Report to the Joint Fire Science Program, JFSP Project #04–4-2–01. Available online: http://citeseerx.ist.psu.edu/viewdoc/download?doi=10.1.1.185.1129&rep=rep1&type=pdf; last accessed 23 May 2018

9 Commentary on science and adaptive management in collaborative restoration

Susan Jane M. Brown

Being a lab rat is an interesting experience. For years, you run your maze looking for your cheese and wonder whether your fellow rodents are running the same maze, looking for the same cheese – or have they figured out a more efficient, successful, and creative way to achieve their ultimate objective? How does one rat learn from another?

Don't get me wrong: I've actually enjoyed being a lab rat – or, more appropriately, a participant in this grand experiment called the Collaborative Forest Landscape Restoration Program (CFLRP). For some years now researchers have been studying CFLRP projects and their participants and, coming from a relatively small pool of individuals, I've spoken at length with most of these researchers at one time or another. Even so, collaborators such as myself are notoriously lacking in our ability to 'tell our stories'. We're doing amazing and important place-based forest restoration, but rarely do we have the time or capacity to explain just exactly what it is that we do to the broader community and stakeholders, and thereby gain their support for our work. Consequently, a volume that collects case studies and stories of collaboration has been overdue, and hopefully this will help fill that storytelling niche.

Seeing one's own experience reflected in the experiences of others creates a sense of camaraderie that is sorely lacking in most federal land management; it is the 'we're in this togetherness' spirit that is a hallmark of successful CFLRP projects across the west. The writings that comprise this section are spot-on in their synthesis of the opportunities and challenges associated with CFLRP and collaborative adaptive management, and present an excellent springboard for decision-makers and stakeholders to improve both the program and the socio-economic and ecological outcomes expected from it. Based on the writings of Toman, Colavito, Cheng, and their teams, as well as my own experience as a founding member of the Southern Blues Restoration Coalition CFLRP, it is my hope that the following lessons learned become institutionalized and operationalized, both within the Forest Service as well as in the communities the agency serves.

As Toman *et al.* write, successful collaboration depends on a shared understanding of the socio-economic and ecological problems faced by large landscapes and the human communities that depend on them. As we have learned with the Blue Mountains Forest Partners (one being half of the

Southern Blues Restoration Coalition, the other half being the Harney County Restoration Collaborative), coming to agreement about what exactly is 'out of whack' on a particular piece of ground, what caused it, and what we can do to fix it, takes what often feels like an inordinate amount of time. However, the Blue Mountains Forest Partners (BMFP) – like many of our CFLRP compatriots – have learned that engaging organizations and individuals who can bring non-partisan scientific information and approaches to the table is key to success. Take, for example, post-fire management: other than old-growth forests, management of recently burned landscapes is perhaps the most electrified third rail of forest policy for many stakeholders. BMFP has been no exception. My timber industry partners have a hard time understanding why I, as a conservationist, 'need all those dead trees for the birds' when there are hundreds of thousands of acres of snag forests across our region: Can't the loggers have at least some of those trees for the mill? And, while I understand the economic interest in quickly harvesting a deteriorating public asset for community economic stability, I also know that historic logging practices have left the overall landscape depauperate in snags, and has resulted in a decline in viability of a number of dependent wildlife species; and my timber partners begrudgingly acknowledge this ecological reality.

What to do? In the wake of the 2015 Canyon Creek fire on the Malheur National Forest, BMFP reached out to our partners in the scientific community and proposed a research experiment: rather than conducting the usual mode of post-fire logging that removes all snags of all diameters from large areas affected by wildfire, what if we designed and implemented an experiment of differing prescriptions designed to achieve the optimal habitat conditions for several woodpecker species at once, while also recouping the economic value of the burned timber? Working with researchers from the Forest Service's Rocky Mountain Research Station, we are now in year three of a four-year experiment to see if we can do the impossible: i.e. come to collaborative agreement over post-fire management. Coupled with existing data collected by researchers from Oregon State University on unburned pre- and post-restoration treatment plots across our CFLRP project area, the woodpecker research should help BMFP develop 'zones of agreement' around post-fire management. Ideally, this joint fact-finding and research could be used and adapted for local conditions in the future by other collaboratives. Win-win-win!

As this example indicates, applied science delivery is critical for the success of collaborative groups seeking to heal old wounds and build new trust and respectful relationships with former opponents, including the Forest Service. However, even with robust scientific engagement from boundary organizations and others involved in science, support for collaboratives and assurances that land managers will implement this learning is often lacking. As Colavito explains, there is often a disconnect between scientific inquiry and delivery on one hand, and decision-making that actually implements either existing scientific information or information that is developed expressly in response to inquiries from collaborative groups on the other hand.

This is a failure of collaborative adaptive management and perhaps one of the greatest frustrations for participants (the other being the nearly constant revolving door of personnel turnover of our agency partners). On the Malheur National Forest, the BMFP worked with forest ecologists and other experts from Oregon State University and University of Washington to develop a scientifically valid and easily implementable method to identify old-growth trees of various species based on the physiological characteristics of the trees rather than a diameter limit that failed to protect old-growth trees and hindered science-based ecological restoration. Our approach, based on the peer-reviewed work of Drs Jerry Franklin, Norm Johnson, and Robert Van Pelt, is incredibly rigorous, informed by extensive locally collected data, and embraced by the BMFP in our 'Upland Forest Restoration Zones of Agreement'. However, we have struggled mightily to ensure that the Forest Service actually utilizes this information in project planning, even though the goal of our approach was to streamline the implementation of restoration treatments that would actually restore the landscape.

Challenges include: adequate environmental analysis (National Environmental Policy Act review), ensuring that our restoration prescriptions are actually implemented as intended on the ground (there is often a disconnect between silviculturists who have been part of the collaborative process and contract administrators who have not, and thus do not understand or support our approach); disagreements between agency personnel who support the stewardship contracting mechanism and those who do not; and the fact that contracts to implement our restoration vision are increasingly awarded to non-local businesses, robbing the local community of much-needed economic development that was a promised outcome from our ten-year stewardship contract implementing our CFLRP project. While there was much initial fervour and fanfare around our CFLRP project and long-term stewardship contract, it sometimes feels as though the commitment to make this new way of doing business work for all stakeholders has waned.

This resistance is odd for our collaborative, since the Forest Service has in fact been supportive of our efforts from the beginning. However, as Cheng and his team observe, there is often an undercurrent of opposition to collaborative adaptive management within the Forest Service for a host of reasons. Consequently, for collaboration to be effective, the 'license to learn' must be embraced by agency leadership and clearly communicated down the line as mission critical. Every agency employee must understand what collaboration is, how it is implemented on the local forest, and why shared stewardship is essential to the Forest Service's ability to achieve its objective of caring for the land and serving people. This leadership starts at the top and must be immune to the political winds that tend to prioritize outputs rather than outcomes. There remains much to do on this front.

While collaboration is often touted as a way for discordant stakeholders to work out their differences amongst themselves, just as often collaboration is a way for stakeholders and the Forest Service to repair broken relationships through what sometimes feels like cage-match marriage counselling. But, collaboration is a two-way street; it is impossible to rebuild relationships with a revolving door. As most authors in this book note, the high turnover of staff in the Forest Service is a substantial barrier to successful collaboration. The culture of 'moving out to move up' may, at one time, have made sense, but in today's challenging political and social climate it is clear that it does much more harm than good. When social license, joint scientific learning, and field experience designed to increase the pace, scale, and quality of forest restoration are compromised by an antiquated personnel policy, it is time to rethink and reform that policy (some might call this *social* adaptive management …).

As a member of a CFLRP project and a local forest collaborative, I would observe that one aspect of successful collaborative adaptive management was not mentioned by the authors of this section: effective socialization. Federal forest management in the west has been characterized by acrimony, distrust, and 'othering' of stakeholders, including federal agency employees who also live in public landed communities alongside non-federal participants. I often worry for my Forest Service partners and understand why some are short-timers without plans to stay in the rural community in which I work. The same can be said for local community members – environmentalists and loggers alike – who take a stand in favour of science-based large landscape restoration. In an era rife with anti-government sentiment, it can be downright dangerous to be a collaborator.

To counter this tense environment, the BMFP has created a culture of socializing with each other around our regular monthly collaborative meetings and field trips. Because the Forest is located off the beaten path and many collaborative members come from outside the county, the group historically has had 'time to kill' in-between meetings; what better way to debrief a difficult conversation about moist mixed conifer management or wildlife viability than over a shared meal or cold adult beverage? Spending unprogrammed time together has allowed our membership to move from adversaries to acquaintances, from acquaintances to partners, and from partners to friends. And like good friends, we don't always agree with each other: indeed, strong disagreements remain common in our collaborative. But what *has* changed is the way in which we disagree, and the way in which we resolve those disagreements. Rather than lobbing lawsuits and personal attacks on each other, we roll up our sleeves and get to work – usually with a lot of field data collection, monitoring, and adaptive management as part of the solution.

This is the kind of real social change that the Collaborative Forest Landscape Restoration Program was designed to encourage. Thankfully, as these authors report, that change is happening in many places, although challenges remain. Still, I remain optimistic that collaborative conservation will free all of us from our mazes.

Part III

Collaborative implementation

10 Collaborative implementation

Implications for adaptive management and restoration

William H. Butler, Ashley Monroe, and Sarah McCaffrey

Introduction

The Collaborative Forest Landscape Restoration Program (CFLRP) set a new standard for mandating collaboration in public lands management. The program required that proposed landscape restoration projects be 'developed and implemented through a collaborative process that includes multiple interested persons representing diverse interests and is transparent and nonexclusive' (16 USC §7303(c)(2)). Moreover, it required a 'multi-party monitoring, evaluation, and accountability process to assess positive and negative ecological, social, and economic effects of projects implementing a selected proposal for not less than fifteen years after project implementation commences' (16 USC §7303(g)(4)). The Act, therefore, set the collaborative requirements for planning, implementation, and monitoring over the life of the projects to achieve restoration goals and undertake adaptive management.

While the US Forest Service has had extensive experience in collaboration, much collaborative practice to date has focused on the planning phase. Partners and collaborators typically have had less of a role as projects moved into the implementation phase. CFLRP challenged this status quo by requiring implementation and monitoring with collaborative groups. At the same time, the agency remains the final authority on what can be done on National Forest System lands. This longstanding context for managing national forests has shaped the professional identity, expectations, and practices of agency personnel for more than a century. The key questions of this chapter are what activities constitute collaborative implementation and what does collaboration through implementation offer to large landscape forest restoration projects?

This chapter sheds light on what activities can be undertaken in collaborative implementation, how these activities shape the potential for adaptive management, and the extent to which collaborative implementation might contribute to increasing the pace, broadening the scale, and enhancing the quality of restoration. We begin the chapter with an overview of the literature on collaborative implementation and adaptive management. Then, we describe the activities that CFLRP participants consider to be contributions to collaborative implementation. We interpret these activities through a framework based on implementation and adaptive management literature. We conclude with implications to theory and practice of collaborative environmental management.

Collaborative implementation and adaptive management

The rise of environmentalism in the 1960s and the resultant environmental policies passed in the 1970s gave rise to two key approaches to environmental governance that have been highly influential in US forest management. First, the legislative foundation built by the National Environmental Policy Act (NEPA) of 1969, the National Forest Management Act of 1976, the Endangered Species Act of 1974, and subsequent rules and regulations contributed to gridlock on the forests and a decline in the timber industry by the 1980s. To find a path forward for active management, erstwhile adversaries began working together in specific places to try to find collaborative solutions to the conflict (Wondolleck and Yaffee, 2000; Luke, 2002). Second, the recognition of failed past management and a growing distrust in government, coupled with a general scepticism of scientific management, contributed to a shared recognition that the uncertainty and complexity of forest ecosystems and how to manage them was much higher than previously thought. This shook the foundations of the scientific management paradigm that had guided forest management (and public administration more generally) since the Progressive Era, and it legitimized the rise of adaptive management which promised a way to handle uncertainty and still undertake responsible management (Cortner and Moote, 1999; Luke, 2002).

Since that time, collaboration in environmental management has been on the rise to the point of ubiquity in forest management in the United States. However, collaborative groups have, most frequently, played a role in the planning phase of projects and have left implementation to the responsible agency with the authority to manage the land. Gray (1989) and Margerum (1999, 2011) point out that effective collaboration through project implementation has been elusive for a variety of reasons, including statutory constraints, organizational or administrative procedures and requirements, new players coming into groups after the planning is complete, new issues that arise during implementation that need resolving, and the inertia of familiar ways of working. Gray (1989) further contends that collaborative groups are particularly susceptible to collapse during implementation and argues that 'if implementation issues are not assiduously anticipated during the negotiation [or planning] phase, implementation is guaranteed to pose new conflicts.' However, if they continue to work together beyond the planning phase, by designing new forums or committees as needed, ensuring organizational or constituent buy-in from diverse participants, and monitoring progress throughout to ensure compliance, then collaborative groups can play an important role in project implementation.

Collaborative implementation often involves a range of actors from multiple sectors (Koontz and Newig, 2014). Thus, authority and responsibility for action may be shared or distributed across multiple actors within the collaborative group. Networks may be important arbiters of what gets implemented and by whom. Margerum (2011) theorizes that social networks can often be helpful in inspiring or guiding actions at a community or individual level; interorganizational networks can be effective at pushing for organizational level changes and actions; and policy networks can effectively promote policy changes to respond to a shared problem. A collaborative group may draw on these different types of networks to affect the change they envision.

Koontz and Newig (2014, p. 420) write more explicitly about collaborative implementation in settings where a government agency has the primary authority for management, noting that '[C]ollaborative implementation commonly involves recommendations without political authority to carry them out.' In their research on watershed management projects, they distinguish between top-down approaches to implementation, which often characterize conventional hierarchical management, and bottom-up approaches, which require understanding the dynamics of network actors who contribute to implementation. They conclude that while each of the watershed partnerships in their study was unique in its own context, the cases with higher levels of collaborative participation in project implementation generally had sufficient resources (funding), leadership (often a full-time coordinator), robust networks across levels, and willing landowners. Even with these insights, the literature on collaborative implementation remains sparse and questions persist. Of importance in this research is whether collaborative implementation can increase the pace, broaden the scale, and enhance the quality of restoration.

Alongside the increased focus on collaborative governance in environmental management, adaptive management has risen to the forefront as the preferred approach in the face of uncertainty. The literature on adaptive management suggests that managing complex and uncertain social-ecological systems requires feedback loops between planning, implementation, and monitoring (Holling, 1978; Lee, 1993). When Holling (1978) conceptualized adaptive management, he emphasized the uncertainties associated with our understanding of biophysical system dynamics and the need for an experimental approach where monitoring helps determine what strategies are most effective. Such an experimental approach focuses on reducing uncertainties through management interventions. A slightly less rigorous take on adaptive management emphasizes resource objectives and enhanced decision-making through monitoring and evaluation, without the explicit focus on experimentation (Williams, 2011). This approach still relies on high-quality and robust monitoring, but usually measuring what was attempted (implementation) and what worked (effectiveness) rather than resolving uncertainties (research or validation) (Egan, 2013). An approach that involves collaborative implementation can be enhanced by feedback through monitoring, and a robust adaptive management strategy permits learning and knowledge building to be integrated into the implementation of projects.

While adaptive management has been widely suggested as an effective way to address problems high in uncertainty, the integration of adaptive management into practice has been more difficult to achieve (Lee, 1999; Allen and Gunderson, 2011). Allen and Gunderson (2011) identify nine pathologies that limit how adaptive management is undertaken, most of which are the result of institutional and management contexts rather than an inability to perform scientific and learning tasks. Key among these is a lack of stakeholder engagement, a consideration that Allen and Gunderson (2011) contend was conceived of as a bridge between managers and scientists, or between science and values.

Collaborative adaptive management brings these ideas together with an emphasis on stakeholder involvement throughout the adaptive management process, not only to ensure buy-in, but also to expand opportunities for learning through diverse perspectives and expertise. The collaborative process can contribute to effective adaptive management as knowledge is shared and generated through deliberation and experimentation, which depends on collaborative engagement throughout planning, implementation, and monitoring (Armitage *et al.*, 2007; Olsson *et al.*, 2007; Pahl-Wostl *et al.*, 2007). While collaboration through an adaptive management process is important, less is known about how collaborative engagement throughout planning, implementation, and monitoring can impact adaptive management processes in large landscape restoration.

Research questions and methods

In this research, we posed two key questions: (1) What do participants in a collaborative process actually do which they consider to be linked to the implementation phase of restoration projects? (2) What are the implications for adaptive management and restoration efforts of engaging in collaboration throughout planning, implementation, and monitoring? We explore these questions through a qualitative case study approach of the first ten landscape collaborative groups selected for CFLRP in 2010. See Esch and Butler in this volume for an overview of these cases. We studied the cases from 2010 to 2016, adding interviews and case documentation to our database throughout that time. We conducted more than 120 interviews (over forty with agency personnel and over eighty with participating stakeholders) at the time of this writing, visited eight of the landscape groups to tour restoration areas and observe collaborative interactions, observed six regional meetings that included several of our case study sites at each meeting, and gathered documents such as NEPA planning documents, meeting minutes, project reports, and proposals for analysis. The focus of the research from the beginning was to understand what constitutes collaborative implementation and what tensions and opportunities arose as participants navigated this largely unfamiliar terrain. We conducted semi-structured interviews and posed questions where respondents were asked to describe what activities they thought collaborative groups were undertaking to contribute to implementation; evaluate how well those activities were going; elaborate on how those activities contributed to implementation specifically; identify what challenges arose between collaborative groups and agency staff; and, generally, assess the progress that CFLRP projects were making and why they were making more or less progress toward restoration goals. After the first year of the research, we conducted follow-up interviews with key informants on most projects to obtain a longitudinal perspective on the cases.

From the beginning of our research, the aim was to build theory grounded in practice about how collaborative implementation is understood, how it is undertaken, and how it interacts with adaptive management processes and restoration outcomes. We used a grounded-theory strategy for analysing the data, an iterative process involving a multi-level coding approach that examines, names, and categorizes the context, conditions, actions, and consequences related to phenomena of interest (Charmaz, 2006). Using an open-source, qualitative coding software, we analysed sixty interview transcripts of CFLRP landscape collaborative participants from interviews conducted in 2011 and early 2012. We refined the coding scheme and adjusted our interview protocols as we continued our interviews and documented analysis over the subsequent four years. This process of modifying and reinterpreting our initial theoretical concepts ensured that the theory could be grounded in and emergent from the research data. We tested our assumptions and interpretations with key informants after developing our interpretive framework to ensure that our categories and explanations resonated with practitioners. These efforts helped us further refine our interpretations, which we present in the sections that follow. Given that we were aiming to construct theory for collaborative implementation and adaptive management in general, we did not conduct comparative case analysis but, rather, built our theoretical framework drawing data from across all cases.

Collaborative implementation in CFLRP

To address our first research question, this section describes three activities that the CFLRP participants we interviewed associated with collaborative implementation: prioritization, treatment enhancements, and multi-party monitoring. Conventional thinking suggests that prioritization is primarily a planning activity, while treatments are linked to implementation, and monitoring is associated with evaluation. However, CFLRP participants had a less compartmentalized view of how these activities overlapped and related to implementation. This section briefly elaborates on each of these activities and clarifies how participants characterized them.

Prioritization

Prioritization in landscape restoration involves deciding where to conduct activities (e.g. the removal of woody biomass or 'fuels'), what strategies to utilize (e.g. prescribed fire or mechanical thinning), and when to undertake the planned work. Priority setting requires determining desired future conditions, what scientific knowledge sheds light on the context, what values are emphasized, what resources are available, and sequential restoration activities.

When discussing prioritization, most CFLRP participants noted that they engaged in dialogue during the CFLRP proposal-development phase or during the development of planning documents required under the National Environmental Policy Act (NEPA). These dialogues often involved discussing landscape priorities using baseline data such as the Ecological Condition Model used in the Accelerating Longleaf project in Florida, or previously agreed-upon restoration principles which informed planning and proposal development in the Colorado Front Range. Many also discussed how they engaged in ongoing prioritization through group deliberations as the project unfolded. According to the interviewees, this continuing dialogue about prioritization represented a more direct link to implementation. Through committees, workgroups, or another decision-making body in the collaborative group, time was dedicated to prioritization efforts on a periodic basis where group members, often working interactively with Forest Service staff, revaluated where to work next and what to do based on assessment of the work that had already been accomplished.

In several cases, participants described how ongoing input into agency decisions allowed them to feel as though they continued to have an influence on what strategies the agency would turn to next. An NGO partner on the Clearwater Basin Collaborative clarified that, 'Ultimately, the forest supervisors are the decision makers. That being said, to date, they have shown a great deal of respect and honouring of our opinion and input into decisions.' On the Tapash, landscape-level planning work identified what the target areas and sub-areas for restoration would be, but an NGO collaborator observed that 'it wasn't outlined in order and how that would work exactly'. Instead, agency staff described how they turned to the collaborative group to be 'involved in determining which is the next landscape' area and what treatments should be attempted, noting that the collaborative 'has pretty significant input in terms of prioritizing'. As the project evolved, the Tapash collaborative set up project-level teams to have a more inclusive process with stakeholders closer to the ground to help inform what restoration strategies should be deployed and where. Prioritization on the Uncompahgre was described by one collaborator as following a similar process: '[W]e will get the lead folks to sit down and explain the different projects. Then, there will be some moving around as they tell us what their priorities are and why, and then we'll see if we agree and then they start tweaking them.' On the Dinkey project, an agency staff member observed that, as they approached an internal deadline for budget allocations, they sought collaborative input on prioritization, asking group members to vote on the highest priority items from a list of projects. That input was then used by the agency to help determine which projects to move to the top of the list.

Ongoing prioritization thus allowed for direct feedback and strategizing about what the most effective restoration treatments and locations might be as implementation work moved across the landscape. It also allowed for the re-evaluation of decisions made in the planning process and the need to adapt to changing conditions on the ground. Wildfires were a common experience across several landscapes. These disturbances dramatically changed ecological conditions and led to the need to reconsider existing prioritization of certain strategies or projects. Working within larger landscape project areas, which created more flexibility in terms of prioritization and location of treatments, and having mechanisms in place for the collaborative to provide input through ongoing prioritization, allowed project teams to more easily adapt to changing conditions.

Treatment enhancement

Interviewees generally identified conducting treatments, such as mechanical thinning, prescribed fire, invasive species removal, and habitat improvements as the central function of implementation and what CFLRP funding was supposed to be used for. Working alongside agency personnel to conduct treatments was desirable for many in the collaborative groups, but the ability to do so was often hindered by policies and rules designed to ensure that the agency is held to account for project implementation. For example, to conduct prescribed fire treatments on federal lands, individuals must obtain certification in nationally established Interagency Standards for Fire and Fire Aviation Operations. Few outside the agency have surmounted this hurdle, including many state forestry staff who, as a result, are not permitted to burn alongside their federal counterparts on federal lands. A notable exception includes staff members from The Nature Conservancy who, in some states, have been able to augment agency capacity with their own certified burn crews. Moreover, collaborative participants are often unclear of the process by which timber contracts are designed and awarded, limiting their ability to provide input on treatment implementation and frustrating some participants when contracts are awarded to non-local businesses (e.g. see Brown, in this volume).

CFLRP collaborative participants have, nonetheless, had meaningful influence in enhancing or complementing landscape treatments in ways that contribute to restoration outcomes. As planning documents inform implementation documents and activities, there is the potential for something to be lost in the translation from plans to implementation on the ground, which is usually conducted by contractors not involved in planning processes. Planning documents for restoration projects are often not highly specific, especially in the context of large-scale restoration where planners focus more on outcomes rather than prescriptive actions, leaving room for interpretation and adaptation. Some collaborative groups are addressing this concern by providing training and guidance for contractors who are hired to perform restoration tasks for the agency. On the Dinkey project, collaborative participants helped develop a guide for staff with specific instructions and examples, including photographs, of potential trees that serve as valuable habitat for Pacific Fishers (*Pekania pennanti*), an endangered species on the landscape. The group has widely distributed the guide to Forest Service personnel and contractors who are in the field marking treatments for thinning projects so as to ensure that trees that could be a viable habitat are not cut down. The Deschutes collaborative has conducted field excursions with contractors and marking crews so that collaborative members can clearly communicate to the crews what they would like to see in terms of restoration treatments and why. One collaborator noted that these field visits 'create shared-learning and training opportunities between contractors, stakeholders, collaborators, and Forest Service staff regarding implementation efficiencies of restoration treatments'. Several participants noted the importance of these efforts; as one collaborator stated, 'Participation of contractors in the collaborative or stakeholder communication to contractors is critical to contractors understanding and accepting more complex implementation prescriptions … allowing them to more effectively implement restoration projects.'

Other groups are engaged in undertaking treatments that are complementary to Forest Service restoration work. Given that the CFLRP funding could only be used to pay for 50 per cent of the costs of implementation and monitoring of the projects on national forest lands (16 USC §7303(f)(1)), agency personnel welcomed any work that could help meet the cost-share requirements of the program. According to agency staff, by the fall of 2012, collaborators on the Southwest Jemez project had completed nine restoration landscape treatments, often working on activities such as wetland or aquatic restoration, which complemented the projects of the agency focused on upland and riparian vegetation management. On the Colorado Front Range, in the interest of protecting valuable drinking water supplies, both Denver Water and Colorado Springs Utilities contributed around $2 million each year to support CFLRP restoration projects on federal or adjacent lands. According to both agency staff and collaborators, these contributions significantly expanded the capacity and pace of restoration work conducted by the Forest Service. When collaborative partners shared resources, more work was able to be accomplished than if the agency had been working on its own. According to the five-year CFLRP report, from 2010–14, more than $210 million worth of cost-share or leveraged funding and in-kind services was contributed by agencies and organizations to accomplish CFLRP related objectives, to augment the $155 million appropriated by Congress to the program during that same time period.

Monitoring

All CFLRP projects have some level of multi-party monitoring as a requirement of the program. Nearly all groups have completed and published monitoring plans, and we found that most of those plans address adaptive management strategies for integrating monitoring data back into implementation strategies. In general, monitoring plans include strategies to assess implementation (did we do what we said we were going to do?), effectiveness (did we get the changes that we were hoping for?) and validation (did the theories or hypotheses or assumptions we were testing hold up?). In practice, and reinforced by interpretation of the Act, CFLRP monitoring has focused on effectiveness monitoring rather than monitoring oriented toward testing scientific hypotheses. Social and economic monitoring is also central as the Act requires that projects contribute to local resource dependent economies.

In many CFLRP groups, scientists and specialists run a systematic monitoring program oriented toward assessing changing ecological conditions in the face of management. Such long-term monitoring programs help to reduce the uncertainties associated with predicted outcomes and contribute to longitudinal analysis of management effectiveness. It is important to note that at the time of our interviews, more than five years into the program, no landscape collaborative had reported completing systematic monitoring to the point where data was being fed back into models to adjust management strategies.

Most CFLRP groups interpreted 'multi-party monitoring' to include not only scientists, but also other collaborative partners. A broad suite of stakeholders tended to be involved in qualitative field reviews, which are oriented to pre- and post-treatment evaluation in the field. These excursions are less systematic and often involve going to a site before and after treatment to see if results matched the project intent. On the Deschutes landscape, a collaborator contended that field excursions had been 'a very real time feedback mechanism' followed by 'a real time discussion after implementation is completed'. The feedback was subsequently incorporated into ongoing planning and implementation decisions by agency staff. Another collaborator noted that, 'These multiparty monitoring field reviews have played a key role in providing feedback on implementation results, and that's written down … and making sure that interdisciplinary team planners hear that.' Moreover, field trips were seen as more than just a way to get out into the woods together and eyeball results. As summarized by several participants, 'We are looking at the vegetation, doing quick counts of stems, checking basal area with our thumbs' – in short, they were conducting a very rapid assessment of the area. While no one we talked to said that this type of field monitoring should replace a more systematic approach, many articulated that this was more than just a tour for show – they were collecting data, albeit informally.

These monitoring field visits were also key opportunities to explore social values in the field and how they manifested in restoration project design and implementation. On the 4FRI landscape, agency staff and collaborators marked pre-treated areas together, reviewed sites post-treatment, and reflected on the visual results. These excursions led to productive and challenging discussions about the acceptable level of thinning that should be undertaken. According to an agency staff member, participants ended up realizing there was actually 'a great deal of agreement within the range of silvicultural prescriptions that could be applied'. On the Uncompahgre, participants went on camping trips together to collect monitoring data and started to grapple with value differences. Stakeholders and agency staff learned that the joint monitoring process did more than just expand the monitoring data available; it also helped build relationships and enabled discussion of different social values and perspectives. Both agency staff and collaborative members of the group described this experience as a 'game changer' as it created a shared understanding and started to forge bonds of trust. On the Deschutes, a stakeholder remarked, 'These field reviews have been incredibly important in promoting shared learning and trust between stakeholders and the Forest Service.' Thus, by working together, through monitoring to evaluate the effectiveness of implementation strategies, collaborative groups contributed both to how implementation reflected their collective understanding of the intersection of science and values, and to laying the groundwork for more effective adaptive management.

These discoveries in the field were repeated in other landscapes where stakeholders and agency personnel, who seemed to disagree with each other around the negotiating table, would soften their positions when faced with project evaluations in the field. Stakeholders and Forest Service personnel across many landscapes indicated that they witnessed perspectives about what constitutes effective restoration shifting during these excursions. While it was not completely clear that field reviews led directly to a change in implementation strategies or actions on the ground, some of our interviewees felt that this shifting of perspectives was a step that could lead to changes in subsequent implementation strategies or actions.

Theoretical implications

To address our second research question regarding contributions to restoration and adaptive management, this section elaborates on two key attributes of collaborative implementation. First, it describes how collaborative implementation can enhance the quality, increase the pace, and broaden the scale of restoration. Second, it explores how the activities of collaborative implementation, described above, link to adaptive management processes. We found that the potential for collaborative implementation to enhance social-ecological restoration and strengthen adaptive management was high. However, it should be clearly noted up front that not all cases engaged in all of these activities, nor did they do so equally well. Our ideas here draw on merging insights from all cases collectively to explore the potential of collaborative implementation to contribute to restoration outcomes and adaptive management, while recognizing that it does not necessarily yield these results in all contexts.

Enhancing restoration

A key question in collaborative environmental management is whether collaboration can enhance the quality, pace, and scale of restoration work. A clear answer to this question remains elusive. However, these findings suggest that there are significant gains to be made by engaging in collaboration throughout all project phases, as collaborative implementation is less rigid than conventional processes of planning, implementation, and monitoring. Rather than getting agreement on a static management plan, collaborative implementation allows for discussion to facilitate stakeholder agreement throughout the life of the project and to leverage shared work. Participants on several landscapes argued that collaboration throughout implementation and monitoring allowed them to move beyond project-level NEPA assessments to more landscape-level NEPA plans. It also enabled restoration efforts to expand beyond federal lands and across landownerships. These findings are corroborated by program-wide survey results as well (see Schultz and MacIntyre in this volume).

One of the critiques levelled against collaboration is that it is time-consuming (Layzer, 2008). However, conventional planning processes can also be time consuming, so this critique may not be relevant in the context of public lands management. In NEPA planning processes, concern about potential litigation can increase the time needed for planning as Forest Service staff seek to make the final document 'litigation proof' (Stern *et al.*, 2014). But, by working through collaborative implementation and building flexibility into the process, most CFLRP projects seemed to have warded off litigation and increased the spatial extent of planning projects. Bringing stakeholders into the process early can provide legitimacy to the outcome, so long as the stakeholders believe their interests and values were incorporated. As of early 2018, the national CFLRP coordinator reported that there had been only one CFLRP project that had faced a challenge under NEPA.

On many projects, stakeholders were willing to support larger project footprints that focused on the landscape level rather than a forest-stand level – if they had assurances of continued engagement and could help the agency adjust strategies going forward. The Colorado Front Range, Uncompahgre, and 4FRI projects all initiated NEPA planning at unprecedented spatial extents under CFLRP, with stakeholders involved at various levels providing input into the process. Other projects under CFLRP also engaged in planning at larger spatial extents than their usual practice. One district ranger shared that she worked with collaborators for more than two years to get agreement on a large-scale NEPA document, saying:

> You can spend two years fighting litigation or, you can spend two years (or a lot of time) upfront developing a proposal that has buy-in from the community … it is a short Environmental Assessment. It gives us a lot of flexibility. We didn't draw polygons and say were going to treat here with this treatment or here with that treatment. We basically just identified a

range of acres that we thought they would treat and a range of different prescriptions and ways we might treat them.

That flexibility was based on social license for the range of proposed activities and allowed them to cover a much larger area under one planning document, adding efficiencies to their planning process; in this way, the added value of the collaborative process can save time down the road by reducing the number of separate NEPA documents needed.

Collaborative implementation does not easily inform larger scale planning efforts in every case, however. It took years of deliberations and frustration on the part of stakeholders involved in the Colorado Front Range project to eventually gain enough traction to become more involved in the NEPA planning processes on projects initiated in the second five years of CFLRP. Similarly, on the 4FRI, stakeholders complained that the collaborative group was ill-informed and not sufficiently involved in the early drafting phases of the largest Environmental Impact Statement written under the CFLRP projects (Butler, 2013). However, some stakeholders and agency staff have pointed to the importance of collaborative group deliberations and recommendations on influencing some of the content of that exceptionally large-scale planning document.

Leveraging shared work also can help improve the pace, scale, and quality of restoration. Under the CFLRP, this was achieved primarily through cost-share opportunities. Those that took advantage of cost-share activities described an increased pace of restoration through funding, or by adding staff to project teams to increase overall restoration capacity. As a result, some groups covered more area more quickly than originally planned, such as in the Accelerating Longleaf and Front Range projects, while others completed non-fuels related restoration projects, such as aquatic restoration work in the SW Crown and SW Jemez or the invasive species reduction and native species reintroduction projects on the Uncompahgre.

A final question is whether the quality of restoration outcomes is improved through collaborative implementation. One of the critiques of collaboration is that agreements sometimes involve compromise and devolution to the lowest common denominator (Coglianese, 1999; Layzer, 2008). Our work indicates that collaborating during the implementation phase tends to expand the possibilities for improving the quality of restoration treatments, as the resources, input, and capabilities of stakeholder groups can increase the capacity to achieve restoration goals. CFLRP collaborative groups helped ensure that planned treatments were translated into prescriptions that aligned with group values by conducting training in the field and developing guides. They communicated directly with contractors and operators in the field about the key values at stake and desired treatment attributes. In some cases, they helped to resolve complex scientific debates by doing joint research and monitoring projects in the field using other resources, such as grant funds, to support their work. These added contributions through collaborative implementation arguably contributed to enhancing the quality of restoration outcomes by clarifying values, resolving scientific debates, and building projects that reflected the range of shared values of collaborative participants. However, we did not conduct an objective scientific analysis of outcomes to determine whether the need for compromise in collaborative deliberations watered down the effectiveness of treatment strategies. Moreover, these outcomes may be perceived as better outcomes according to those involved in the program, but that does not address the concerns of those who felt excluded or unable to have their voices heard in the process. Thus, we must conclude that, while the potential for increasing the pace, scale and quality of restoration is increased through collaborative implementation, our findings cannot provide a definitive answer on these questions.

Adaptive management

Overall, we found that collaboration through implementation and monitoring can increase the potential for effective adaptive management by creating mechanisms for more consistent feedback, which can then be integrated into management decisions. Figure 10.1 depicts how feedback loops from collaborative efforts can add nuance to the conventional adaptive management model.

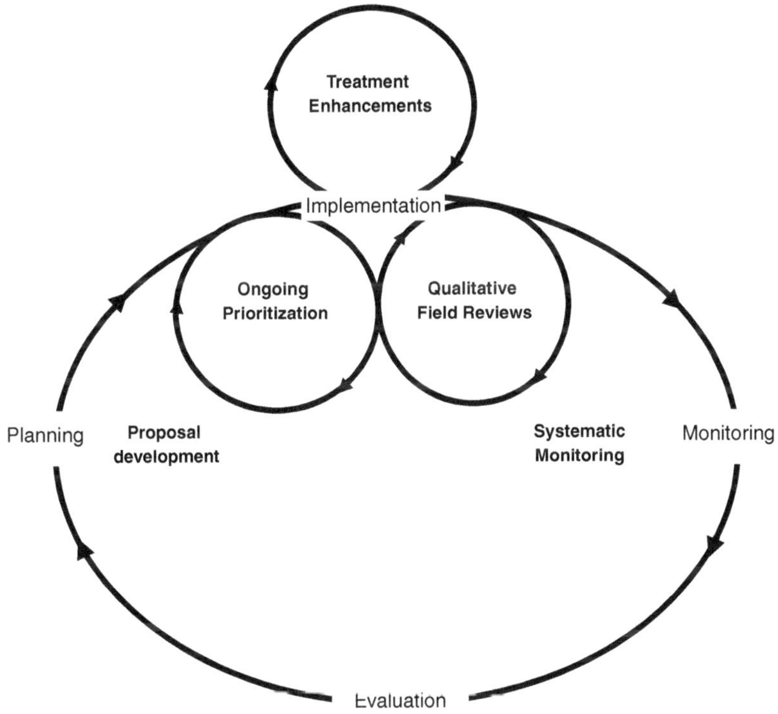

Figure 10.1 Adaptive Management Cycle in Collaborative Implementation

Source: author-created (adapted from Butler, Monroe, and McCaffrey, 2015).

Proposal development laid the foundation for landscape-scale management by providing a strategic plan based on restoration principles and ten years of projects to be completed on the landscape. As the CFLRP landscapes shifted toward implementation, three interconnected feedback loops emerged. First, ongoing prioritization allowed stakeholders and agency staff to discuss how things were going and what should be done next. This loop has the potential to influence the design and implementation of new projects, or recalibration of existing projects. Second, treatment enhancement activities ensured that restoration followed agreed-upon strategies and priorities through training and guidance tools, or augmented treatments through cost-share and collaborative work. This loop has the potential to increase the pace, scale, and quality of restoration strategies on the ground through added capacity and improved alignment with stakeholder values. Third, qualitative field reviews helped inform ongoing prioritization work and implementation activities by providing real time feedback to land managers, who could then alter their strategies within a sub-landscape project area with similar characteristics. These feedbacks can inform projects within the designated landscape area, allowing groups to adjust treatment strategies and locations within approved planning areas or to adjust NEPA level planning documents from project to project. Systematic monitoring is a long-term endeavour that also has the potential to inform further planning activities. However, the time lag between data collection, analysis, and interpretation in relation to implementation means that results may not come in until the projects are complete. These monitoring efforts are more likely to inform strategies in the next landscape rather than within current landscape boundaries.

Collaborative implementation thus appears to encourage adaptive management and offer the potential for more rapid and frequent feedbacks. Systematic monitoring is necessary to demonstrate that the desired effects have been achieved through treatments on the landscape. However, qualitative field reviews, ongoing prioritization, and activities designed to enhance treatments can more readily consider changing landscape conditions, clarify and refine social values, and incorporate learning that occurs through review of management actions. Rapid feedback occurs through dialogue and engagement with stakeholders and contributes 'real time' input in the field. This has the potential to influence new strategies that can enhance implementation efficiencies and effectiveness.

This adaptive management approach is exemplified by the advantages of qualitative field reviews undertaken on many of the landscapes. On the Deschutes landscape, qualitative field reviews gave the Forest Service the flexibility to attempt 'designation by prescription', or allowing contractors to undertake thinning projects without marking specific trees on the treatment area first. According to multiple collaborators, partners felt comfortable leaving the decision-making in the hands of the contractors when they had been to the field with them, helped run some trainings, and knew they would be able to review the results and make adjustments. The reliability of a rapid feedback mechanism proved to be essential and led to significant time and cost savings in project implementation.

The Accelerating Longleaf project provides another example of these dynamics. After initial treatments had been completed, a major fire event created a need to identify new treatments in an area of the landscape that originally had not been on the CFLRP work plan for the first few years. This change impacted the systematic monitoring plan developed by the Tall Timbers Research Station, as there were no treatments conducted in the plots they had laid out for monitoring data collection for the first few years of treatments. Meanwhile, Forest Service staff took partners on field excursions to see examples of treatments they had conducted and to get feedback on what seemed to be working well and what was not working well. As a staff member shared, 'We write down these notes and bring them back to our interdisciplinary team meeting the very next week.' This informal feedback mechanism was also used by the staff in the field as they took photographs and notes on their regular visits to evaluate project outcomes. These qualitative field reviews provided a rapid feedback mechanism and allowed the staff to test different treatment combinations to determine the most effective approaches for restoration, given the abundance of understorey fuels in the landscape. They did not have to wait for the systematic monitoring data to make adjustments to their treatment strategies. If they had, they would have waited much longer than originally planned due to the impact of a wildfire that altered landscape conditions and changed the priority locations to areas that were not scheduled for monitoring in the original plan.

Policy and practice implications

Based on our findings in this research, there are several key implications for policy and practice. First, collaboration through the planning, implementation, and monitoring phases of a project can help address several important aspects of effective restoration. Although the degree that collaboration enhances outcomes remains unresolved, we found that collaborative implementation contributed to increasing the pace, scale, and quality of restoration in rather substantive ways. Thus, it may be worth the investment of time and resources to engage in these collaborative processes on large-scale and long-term projects, given the potential value added.

Second, collaborative implementation, as experienced in CFLRP, suggests how adaptation can be more effectively incorporated into projects over time and at multiple scales. Adaptive management feedback can be more rapid and responsive to changing ecological conditions, as well as changing social conditions, perspectives, and values. While management decisions are informed by science, they are rarely made solely based on science (no matter how good the data). Ultimately, social values are a key determinant in policy and management. Thus, collaborating through implementation and monitoring creates opportunities to clarify and refine these values, so as to respond to changing values as implementation unfolds. Over the past four decades, policy has increasingly encouraged public participation and collaboration in planning. Our results suggest that integrating collaboration throughout all phases of project planning, implementation, and monitoring could further improve the ability to take social values into account in management efforts.

Third, we found that adaptive management on large landscape forest restoration can be enhanced through multi-party monitoring that is both qualitative and systematic. The use of qualitative field reviews allowed stakeholders, agency staff, and implementers to be in the field together, to review projects and provide real time feedback, and to address and clarify social values that underpin treatment prescriptions. These field reviews, while not a replacement for systematic monitoring, could be encouraged as a complementary and more direct feedback mechanism in the monitoring program. It would be an interesting exercise to calibrate these field excursions with more systematic efforts; not to replace the systematic approach, but to see how closely the two align. In order to maximize learning opportunities, for effective field reviews it is important to ensure that members have the opportunity to look not only at successes, but to look also at mistakes or treatments where the desired outcomes were not achieved.

The challenges to engaging in collaborative implementation on federal public lands management are not easily overcome, especially since the agency cannot relinquish authority or responsibility for decision-making. Despite these limitations, this research shows that collaboration through implementation and monitoring is not only feasible, but has the potential to contribute to ongoing restoration efforts enhancing the quality, pace, and scale of treatments while also informing prioritization across the target landscape. Such outcomes suggest that even with the challenges and the time and resources involved, collaboration that continues through implementation and monitoring is a worthwhile undertaking that can build social license, encourage scaling up, and increase the pace and improve the quality of restoration on public lands.

References

Allen, C. and Gunderson, L. (2011) 'Pathology and failure in the design and implementation of adaptive management', *Journal of Environmental Management*, 92(5): 1379–1384

Armitage, D., Berkes, F., and Doubleday, N. (2007) *Adaptive co-management: collaboration, learning and multi-level governance*. Vancouver, Canada: UBC Press

Butler, W., Monroe, A., and McCaffrey, S. (2015) 'Collaborative Implementation for Ecological Restoration on US Public Lands: Implications for Legal Context, Accountability and Adaptive Management', *Environmental Management*, 55(3): 564–577

Charmaz, K. (2006) *Constructing Grounded Theory: A Practical Guide Through Qualitative Analysis*. Thousand Oaks, CA: SAGE Publications

Coglianese, C. (1999) 'The Limits of Consensus: The Environmental Protection System in Transition: Toward a More Desirable Future', *Environment: Science and Policy for Sustainable Development*, 41(3): 28–33

Cortner, H. and Moote, A. (1999) *The Politics of Ecosystem Management*. Washington DC: Island Press

Egan, D. (2013) 'Multiparty Monitoring', In Egan, D. and Dubay, T. (Eds.) *Breaking Barriers, Building Bridges: Collaborative Forest Landscape Restoration Program Handbook*. Ecological Restoration Institute of Northern Arizona University, Flagstaff, AZ.

Gray, B. (1989) *Collaborating: Finding Common Ground for Multiparty Problems*. San Francisco, CA: Jossey-Bass Publishers

Holling, C.S. (1978) *Adaptive environmental assessment and management*. New York: John Wiley and Sons

Koontz, T.M. and Newig, J. (2014) 'From Planning to Implementation: Top-Down and Bottom-Up Approaches for Collaborative Watershed Management', *Policy Studies Journal*, 42: 416–442

Layzer, J. (2008) *Natural Experiments: Ecosystem-based Management and the Environment*. Cambridge, MA: MIT Press

Lee, K.N. (1993) *Compass and gyroscope: integrating science and politics for the environment*. Washington DC: Island Press

Lee, K.N. (1999) 'Appraising adaptive management', *Conservation Ecology*, 3(2): www.ecologyandsociety.org/vol.3/iss2/art3/, accessed 21 May 2018

Luke, T. (2002) 'The Practices of Adaptive and Collaborative Environmental Management: A Critique', *Capitalism Nature Socialism*, 13(4): 1–22

Margerum, R. (1999) 'Getting past yes: From capital creation to action', *Journal of the American Planning Association*, 65(2): 181–192

Margerum, R. (2011) *Beyond Consensus: Improving Collaborative Planning and Management*. Cambridge, MA: MIT Press

Olsson, P., Folke, C., Galaz, V., Hahn, T., and Schultz, L. (2007) 'Enhancing the Fit through Adaptive Co-management: Creating and Maintaining Bridging Functions for Matching Scales in the Kristianstads Vattenrike Biosphere Reserve, Sweden', *Ecology and Society*, 12(1) www.ecologyandsociety.org/vol.12/iss1/art28/, accessed 21 May 2018

Pahl-Wostl, C., Sendzimir, J., Jeffrey, P., Aerts, J., Berkamp, G., and Cross, K. (2007) 'Managing change toward adaptive water management through social learning', *Ecology and Society*, 12(2) www.ecologyandsociety.org/vol.12/iss2/art30/, accessed 21 May 2018

Stern, M.J., Martin, C.A., Predmore, S.A., and Morse, W.C. (2014) 'Risk Tradeoffs in Adaptive Ecosystem Management: The Case of the US Forest Service', *Environmental Management*, 53(6): 1095–1108

Williams, B. (2011) 'Passive and Active Adaptive Management: Approaches and an example', *Journal of Environmental Management*, 92(5): 1369–1378

11 Participating in collaborative implementation

The role of collaborative history and context

Hannah A. Bergemann, Courtney A. Schultz, and Antony S. Cheng

Introduction

The CFLRP was the first policy applicable to US national forests, not just to encourage collaboration but to require it throughout the life of its projects. Specifically, the law said Congress would select and fund projects that are 'developed and implemented through a collaborative process' that is transparent and inclusive of people representing diverse interests (16 USC §7303(b)(2)). These requirements for collaboration were not any more specific, making it unclear what collaborative implementation would look like and how it would occur. The Forest Service had prior experience with collaborative planning, but collaborative implementation represented a new policy goal for which there was little guidance or precedent (Butler *et al.*, 2015).

In this chapter, we explore how collaborative implementation, as envisioned in CFLRP, manifested in practice. For two CFLRP projects, we discuss how stakeholders and Forest Service personnel collaborated during the implementation process, the extent to which they were satisfied with their involvement, and the factors that influenced the ability of stakeholders to participate in this process. Our work is based on, and aims to contribute to, theories on how policies are implemented and customized in local contexts.

Implementation of a collaborative policy mandate

The implementation of any policy goal is affected by factors at multiple levels of governance. National laws, federal agency policies, and other agency norms direct the actions of agency staff, who act as policy agents in the field. At the same time, individual agency personnel influence how policies are implemented, often creating variability in policy outcomes due to the preferences of individuals and the environments in which they operate (Lipsky, 2010). The local external context, such as social and economic conditions, also affects the translation of policies at the local level.

In the public policy and administration literature, scholars have developed frameworks that capture these various factors that affect policy implementation and explain variability across sites and contexts. 'Top-down' or 'structural factors' focus on the role of agency-wide institutions, such as the statutory language of the policy and its requirements, agency incentives, and other aspects of agency-wide leadership and communication (Sabatier, 1986; Steelman, 2010). These conditions affect how policies are translated into practice and the extent to which they lead to changes in organizational behaviour (Fernandez and Rainey, 2006). At the same time, local policy agents, or 'street-level bureaucrats', shape policy implementation at the local level, based on their personal motivations, local agency culture and priorities, and local political pressures (Lipsky, 2010). 'Bottom-up' or 'individual' factors in policy implementation models focus on the interaction between street-level bureaucrats and the local setting where implementation occurs; theories propose that the ability of local implementers to adapt to contextual factors is a key determinant of success (Maynard-Moody *et al.*, 1990). Moseley and Charnley (2014) further develop these ideas in their work which examines how individual agency field units (e.g. individual national forests) respond to policy changes. They delineate the importance of local biophysical, socio-political, and economic factors in influencing policy implementation at the local level.

Given this wide array of factors, some scholars examine processes of *bricolage*, whereby local actors combine and reshape existing institutions (i.e. the formal and informal norms that shape practice) based on a local 'logic of practice' (De Koning and Bennecker, 2013). Others discuss this type of activity as 'creative syncretism', whereby actors creatively navigate and shape institutions as they participate in governance and other social activities (Berk and Galvan, 2009). As Cleaver (2001) explains, formal institutions, including new policies, are introduced into a patchwork of other formal and informal institutions that are affected by history, relationships, and local norms of practice that define how they will be adopted and defined in different locations. For these reasons – in addition to the structural and individual variables – local factors (including local ecology, local social and economic conditions, and local practices) are also relevant for how policies are implemented at a local level.

We drew upon this literature to help us understand the factors that directly affected collaborators' involvement and satisfaction with the CFLRP's mandate to undertake collaborative project implementation. We expected structural and individual factors to affect collaborative implementation under the CFLRP. Local ecological and economic conditions were also likely to be relevant. We were also interested in the role of local history and practice, and therefore focused on local collaborative history as a variable for exploration in our work. As a note on language, the policy implementation literature informs understanding on how policy is implemented. In this case, the policy we are considering is the CFLRP, specifically the requirement to collaboratively implement projects. Throughout the chapter, we refer to this as a 'collaborative implementation' mandate.

The changing landscape of collaboration

Since the Progressive Era, an expert-based management approach, grounded in the paradigm of scientific management, drove decision-making in land management agencies, including the US Forest Service (Lachapelle *et al.*, 2003; Sousa and Klyza, 2007). However, particularly in the 1990s, a desire for greater stakeholder participation and a search for alternative approaches to dealing with the widespread conflict in public lands management led to experimentation with collaborative governance (Ansell and Gash, 2008; Sousa and Klyza, 2007; Williams *et al.*, 2013). By the turn of the twentieth century, collaboration had become a more common approach in forest management (Cromley, 2005; Nie and Fiebig, 2010).

However, even when agency personnel pursued collaboration in earnest, efforts were often constrained by the broader agency culture and practice, and inflexible administrative and procedural requirements (Wondolleck and Yaffee, 2000; Butler, 2013; and see the Introduction to this volume). Laws like the National Environmental Policy Act of 1969 (NEPA) expanded avenues for public engagement, but also limited the role of the public to notice-and-comment participation (Wondolleck and Yaffee, 2000). In the early years of the CFLRP, groups struggled to navigate the requirements of the Federal Advisory Committee Act, another law designed to protect the public interest (Butler, 2013). While these laws were meant to increase public oversight and transparency and limit the role of special interest groups, they also were immediately criticized for limiting the participatory space for place-based collaborators in land management (Sousa and Klyza, 2007). The enduring policy question in US national forest management has been how to create formal opportunities for collaboration within existing institutional and practical contexts.

As noted in the Introduction, the need for fuels management and ecological restoration to occur on larger scales prompted communities across the country to form collaborative groups organized around forestry projects (Schultz *et al.*, 2012). The CFLRP emerged in this context, largely because of stakeholder discussions with members of Congress (see Foreword). The law was innovative in that it formally required collaboration throughout the life of projects, from proposal writing to monitoring, but it left open the question of how this would occur, and even whether it would occur, given that there were no clear accountability mechanisms for partners built into the law's requirements. Although collaborative planning and dialogue had become more common across the National Forest System, the CFLRP's requirement for collaboration to occur during the implementation stage of the project was unique among previous collaborative initiatives. In prior activities, collaboration would occur during the planning and, sometimes, monitoring stages, and implementation would be undertaken solely by the agency (Butler *et al.*, 2015). As such, collaborative implementation represented uncharted territory.

Research objectives and methods

Our central question was how the CFLRP collaborative implementation requirement would be implemented at local levels and what factors might drive variability in implementation. Based on the extant scholarship, we presumed that local implementation would depend largely on a variety of factors, including: local implementers; local social, political, economic, and ecological conditions; and local practice, particularly around collaborative planning of forest management projects. Given the decades of experience working collaboratively in some locations, there was reason to believe that the intent of the CFLRP collaborative implementation mandate might intersect well with established collaborative practices, despite the presence of other formal institutions that could pose barriers.

Our primary objectives in this chapter were to: (1) Explore how CFLRP project personnel and stakeholders collaborated through implementation, and whether they were satisfied with their experiences, and (2) Investigate what factors affected the stakeholders' ability to engage in collaborative implementation, as required by law. The two CFLRP projects that served as the focus of this study were the Front Range Collaborative Forest Landscape Restoration Project (Front Range Project) and the Uncompahgre Plateau Collaborative Restoration Project (Uncompahgre Plateau Project).

The Front Range Project consisted of an 800,000 acre landscape along Colorado's Front Range and included the Arapaho-Roosevelt and the Pike and San-Isabel National Forests. The collaborative group associated with this project was called the Front Range Roundtable, and the subgroup that focused on the CFLRP project was called the 'Landscape Restoration Team'. The Uncompahgre Plateau Project spanned a 572,000 acre landscape across the western slopes of Colorado, including parts of the Grand-Mesa, Uncompahgre, and Gunnison National Forests. Although a loose network of groups was associated with this project, the primary collaborative entity was called the Uncompahgre Plateau Partnership.

In 2017, replicating data collection that had been completed approximately every three years since 2010, we conducted twenty-seven interviews with collaborative participants from both of the CFLRP projects, including external partners and Forest Service personnel who engaged with the collaborative groups (numbers associated with individual interviewee quotes are included in the sections below). This longitudinal work was done so as to understand the development of collaborative processes on these two cases over time. Our deep knowledge and involvement with these two cases made them ideal for understanding how local history and experience with collaboration might create a local logic of practice that would then affect how collaborative implementation under the CFLRP occurred. Interviewees included seventeen individuals associated with the Front Range CFLRP and ten individuals associated with the Uncompahgre Plateau CFLRP. We used a purposive sampling (pre-selected individuals) and a network sampling (individuals recommended by interviewees) approach to select individuals for interviews, until a saturation of information on our research questions was reached (Singleton and Straits, 1999). We selected interviewees from meeting attendance forms provided by each collaborative group's facilitators, and interviewed a broad variety of stakeholder groups so that perspectives from the different interests participating in the collaborative groups were represented. The interests represented in the interviews included environmental organizations, the US Forest Service and other federal agencies, local and state agencies, and private industry.

We also observed and participated in collaborative group meetings and site visits between August 2015 and August 2016. Using a participant-observer approach allowed us to gain richer detail regarding the experiences of the collaborative participants and helped us to develop a rapport with the interviewees (Creswell, 2009). This rapport may have helped collaborative participants feel more comfortable sharing some of the negative aspects of their experiences in interviews. In August 2016, we also joined Forest Service Washington Office staff on site visits to both Colorado CFLRP projects. Over a span of five days, we transcribed and recorded discussions between Forest Service staff and collaborative participants related to the projects.

We used this data to continue building case histories of collaboration and to specifically ask about the collaborative implementation process, barriers to participation in implementation, successes, and satisfaction with their ability to influence the implementation process. Following data collection, we coded interview transcripts and observation data in two rounds using Nvivo software. Coding involved organizing the interview transcripts into segments of text that were categorized based on codes (Creswell, 2009). We approached the analysis of the data with some pre-determined codes related to our research questions, but also identified new patterns that emerged throughout the analysis process. Some of the pre-determined codes included 'top-down' and 'bottom-up' to represent the two policy implementation approaches discussed in the literature review, which were later renamed 'structural' and 'individual' so as to maintain consistency with language used in Steelman (2010). New themes and associated codes that were identified included, 'collaborative characteristics', 'local economic conditions', and 'biophysical conditions' to characterize external factors influencing policy implementation.

Collaborative history and implementation of the projects

The Front Range Collaborative Forest Landscape Restoration Project (Front Range Project)

In the aftermath of the 2002 Hayman Fire, the (then) Regional Forester for the Rocky Mountain Region (Region 2) of the Forest Service convened an interagency partnership composed of the Bureau of Land Management, the

National Park Service, and the Colorado State Forest Service to coordinate funding and implementation of fuels-reduction projects on lands under their jurisdiction. Initially called the Front Range Fuel Treatment Partnership, the name evolved to become the Front Range Roundtable after the government-only group expanded to include non-governmental organizations and local governments – at the behest of stakeholders who expressed concern that the agencies were claiming 'collaboration' in rhetoric, but were not collaborating in practice. Even after this expansion, the Roundtable operated more as an information-sharing network than as a group where stakeholders participated in the conception, design, monitoring, and adaptive management of forestry projects, despite efforts to change this course on several occasions prior to the Front Range Project (Cheng *et al.*, 2016). The original structuring of the Roundtable would have implications for the collaborative implementation of the Front Range Project.

With the initiation of the Front Range Project, the Landscape Restoration Team formed as a working group of the Front Range Roundtable to address collaborative implementation of the CFLRP project. Participation in the implementation process was described by interviewees as occurring mostly through informal and indirect processes, such as feedback provided during field trips and Landscape Restoration Team meetings, as well as through participation in monitoring. According to most interviewees, informal feedback from collaborators, during field trips and group meetings, was used to some extent to adjust and inform prescriptions and treatment locations. One interviewee (#9) from a partner agency described an instance where they witnessed project adjustments being made immediately before implementation occurred, stating:

> The logging machinery was in the area working, but trees were being re-marked and units were being redesigned right up to the moment of the contractors entering that part of the woods. It was interesting and exciting to see some of the changes that were being made that quickly.

However, while many interviewees reported that treatments had been adjusted in response to collaborative input, several interviewees expressed that some of their recommendations were not accommodated. Another interviewee (#3) from a partner agency stated:

> [The Forest Service] tried to accommodate most of those recommendations as they can, but there are certain recommendations that they can't achieve. One of them is a size of an opening, another one would be prescribed fire … we just haven't seen that happen as much, and that comes with a whole can of worms as well.

As we discuss more in the next section, some attributed this to local agency staff willingness to work collaboratively, while others added that many projects were already planned prior to the advent of the Front Range CFLRP project.

On this project, there were also attempts to formalize collaborative implementation. The Landscape Restoration Team developed an adaptive management process for the Front Range Project by which the monitoring conducted for the project would be evaluated on an annual basis and used to inform adjustments to treatments for the following year (Aplet *et al.*, 2014 and see Cheng *et al.* in this volume for more details). This adaptive management process was cited by several interviewees as a potential mechanism for the collaborative group to influence the implementation process, and, for some, the development of this process provided a clearer role for collaborators in the implementation phase as the project went on. However, many said the feedback loop between monitoring information and changed approaches on subsequent projects was ambiguous, limited, and lacked clear agency commitment (see Cheng *et al.* in this volume for more on this process). One NGO partner (#6) explained that this was a challenge for the group, stating:

> We definitely struggle with closing the loop on adaptive management and really figuring out what it looks like beyond the box and arrow diagrams … and how it actually works in a social setting, how the recommendations are delivered to the Forest Service for implementation, [and] how they're actually acted upon.

Interviewees provided a mix of both positive and negative responses regarding their satisfaction with the collaborative group's involvement in the implementation of the project. Some participants reported that they had expected the Forest Service to incorporate more of the group's feedback into treatments and to design projects together. Another person (#8) from a partner agency stated:

> I took seriously the line in [the legislation] that talked about collaborative development and implementation of projects. I thought that there would be more actual rolling up the sleeves, sitting down at the table and designing projects with the Forest Service than there has been. That's been a disappointment to me.

Conversely, several interviewees reported satisfaction with their involvement in implementation, including one partner agency interviewee (#9) who said, 'I think it's pretty impressive and positive that the Forest Service has included this big messy element of collaboration [and] stakeholder input into its planning and implementation of the restoration work'.

The Uncompahgre Plateau Collaborative Restoration Project (Uncompahgre Project)

Unlike the Front Range Roundtable, the origin of the Uncompahgre Plateau Project was not a federal-state government partnership, but a community-based initiative. In 1993, the Public Lands Partnership was established by local elected officials and community residents of Delta and Montrose counties to stave off further conflicts between resource users, environmentalists, and federal land management agencies (Cheng, 2006). Since that time, the Public Lands Partnership has initiated several collaborative implementation projects, most notably one led by the Uncompahgre Plateau Partnership, which engaged federal, state, and local governments; non-governmental organizations; and local businesses in on-the-ground ecological restoration projects. This history of community-driven, collaborative implementation of restoration projects laid the groundwork for the Uncompahgre Plateau Project (Cheng and Randall-Parker, 2017).

Comporting with the grassroots, community-led nature of collaboration in this area, there was no formal working group associated with the implementation of the Uncompahgre Plateau Project. An ad-hoc group of stakeholders, working with university scientists and the Forest Service, developed their own reconstruction of historic forest structure on the Uncompahgre Plateau to inform forest restoration prescriptions (Colorado Forest Restoration Institute, 2008). The Forest Service incorporated this information into the NEPA processes and decisions for the projects comprising the Uncompahgre Plateau Project, and stakeholders felt confident that the Forest Service would implement them accordingly. The stakeholders also were heavily involved in the planning processes for two other projects, the Unc Mesa and Escalante projects, and in jointly developing restoration principles to guide the projects. Some interviewees discussed how stakeholder involvement in the planning process and in developing restoration principles helped to ensure that their vision for the landscape would be carried into implementation. One NGO partner (#27) said, '[The stakeholders] feel like the general principles are being translated into actions on the ground.'

Despite the strong participation of stakeholders in planning efforts, several interviewees reported that there was less involvement from stakeholders as the project transitioned into implementation. When asked about the collaborative group's role in implementation, one NGO interviewee (#17) explained, 'There's not a strong role for the collaborative except to check and see if it's going the way that they thought it was going to be going.' Most emphasized their role in implementation as occurring via annual field trips and stakeholder meetings in which Forest Service personnel reported on that year's accomplishments and discussed the treatments proposed for the following year. Several interviewees reported that the feedback provided during field trips led to additional adjustments in treatment prescriptions. To illustrate this, one Forest Service employee (#23) stated:

> There's been a few places where people have noted.... 'Oh, the trees are spaced a little more evenly than we like, so maybe think about doing a little bit more clumping.' Although, it wasn't anything major, we just talked to the marking crew about that.

Interviewees were overwhelmingly satisfied with the collaborative group's level of involvement in the project and in the implementation process. When asked about implementation, almost everyone emphasized their history in planning as being fundamental to explaining their satisfaction, with only informal involvement during implementation. One Forest Service staff member (#21) noted that further involvement may, in fact, be counterproductive, stating: 'It's great to get the feedback and make sure everybody's on board with what you're doing, but if you get too much involvement, then you could start getting nothing done because you're spending your whole time communicating.' That member continued, 'I think the level of involvement has been good.'

Factors affecting collaborative implementation

Agency-wide structural factors

Required environmental analysis of planned actions under the NEPA, and associated decisions based on those analyses, was a major top-down factor discussed by interviewees for both projects. For the Front Range Project, because stakeholders had not been involved in the planning process, the NEPA requirements to implement projects based on formal decisions impeded collaborative participation in implementation. A large portion of implemented acres from the Front Range Project were the result of NEPA 'shelf-stock' (i.e. projects that have final decisions in place and are ready to be implemented) that had been completed prior to the beginning of the CFLRP project. According to the interviewees, this meant that the prescriptions implemented in these areas were constrained by the requirements defined in the NEPA document and associated decision. One Forest Service staff member (#10) explained:

> Where we didn't have flexibility was in the size of openings. NEPA decisions limited the size of openings to a quarter acre, and so when … the collaborative group actually wanted openings of an acre to 5 acres to 10 acres in size, we couldn't make that happen.

Treatment areas covered by shelf-stock NEPA had not incorporated the Front Range collaborative group's input, at least not beyond recommendations that had been offered by the Front Range Roundtable prior to 2010. By contrast, the Uncompahgre Project, where stakeholders had been engaged in planning with the Forest Service prior to the passage of the CFLRP and during the design of CFLRP projects, NEPA-related requirements were viewed as an avenue for participation. Given these constraints, and the opportunity to build collaboratively agreed-upon parameters into NEPA decisions, the history of collaboration around NEPA affected collaborative implementation and satisfaction thereto on the Uncompahgre Project.

Another top-down factor identified by interviewees, particularly on the Front Range Project, was the lack of clarity in the statutory language related to the requirements for collaborative implementation. Stakeholders were not all in agreement on this point, but some argued that it was problematic that the law did not specifically outline a formal role for collaborators and left the requirements vague. One NGO interviewee (#14) proposed that the statute needed to 'Be more explicit about what is meant by collaborative implementation, because just including that term in the legislation didn't get it done'. Furthermore, some stakeholders pointed to an agency culture, which many talked about as being agency-wide, that made the Forest Service slow to change how they approached these projects. Here again, however, perceptions of this were quite different across the two projects, which necessitates a deeper dive into case-specific factors that explain the variability we saw.

The effect of individual agency implementers

Participants on both projects discussed the role of key Forest Service line officers (i.e. district rangers and forest supervisors who are the individuals who possess formal decision-making power in the agency) in facilitating collaborative implementation. Forest Service personnel associated with the Uncompahgre Plateau project were, generally, viewed positively and as pivotal, due to their role in fostering productive collaboration with stakeholders. Conversely, the Front Range Project participants said they experienced local implementers who both facilitated the collaborative implementation process and impeded it. Several participants on the Front Range Project discussed how the staff on the two Forests involved in the project possessed different attitudes towards collaboration. One NGO interviewee (#14) stated, 'The individuals on the [one forest] were more attuned to collaboration, more attuned to ecosystem management, trying to do things differently than we'd done before in order to achieve ecological gains. On the [other forest], much less so.'

External factors: collaborative characteristics and other exogenous influences

Many interviewees pointed to the importance of the characteristics and history of the collaborative groups associated with the CFLRP projects. Stakeholders participating in the Uncompahgre Project had been discussing restoration on the Uncompahgre Plateau with one another and with their Forest Service counterparts for over a decade. There was a great deal of trust built with the Forest Service during the planning process that carried through to implementation. This point was reiterated many times throughout interviews with participants. 'We already had years of working together as a collaborative by the time CFLRP came along, so we had already gotten over the rough patches; we'd already learned to trust the Forest Service and each other,' stated one NGO interviewee (#15). While participants of the Front Range Project did not explicitly discuss trust, the responses of several interviewees indicated some degree of distrust with the Forest Service to carry out their recommendations. Their lack of prior involvement in the planning process came up as an explanation for why they felt the need to be involved in implementation more closely, and yet they felt constrained in doing so effectively.

We also assessed collaborative capacity regarding the agency and group members' technical expertise to participate meaningfully in implementation, as well as their funding and time to participate. Stakeholders associated with the Uncompahgre Plateau Project expressed that there was, at times, limited capacity on the part of the Forest Service to participate in collaboration, which to our knowledge is largely an internal structural challenge for the agency. One NGO interviewee (#15) stated, 'The Forest Service people just didn't have as much time as you wished they would due to people [who] were retiring or maybe being understaffed, people working two jobs.' A lack of capacity on the collaborative group's side was also noted, as participation from stakeholders has waned in recent years.

Other external factors, such as disturbance of the landscape and local economic variables, affected collaborative implementation to some extent. A challenge identified by interviewees was the translation of the collaborative group's project design criteria and objectives to the contractors who conducted the implementation of the treatments. One interviewee (#9) on the Front Range Project, from a partner agency, relayed that they had been told by Forest Service personnel that, 'It's very difficult to explain instructions in the operating guidelines for the people that actually do the treatments from the machines on the ground.' This difficulty in translating prescriptions is linked to the capacity of operators, which can be viewed as a function of local industry capacity and expertise. Additionally, the biophysical conditions of treatment areas also limited where work could be performed. One NGO interviewee (#14) on the Front Range Project mentioned an instance where the collaborative group's recommendation on treatment location could not be accommodated, stating:

> The location of the treatment area was predetermined by operability constraints and [the Forest Service] basically said, 'The only place that we can harvest timber is on the flat part of the mountain at the top,' even though all the analysis showed that all the values at risk were down at the bottom of the hill where operability was severely constrained.

Comparing collaborative implementation between the cases

Our questions focused on how the CFLRP's collaborative implementation mandate was manifested on these projects, whether stakeholders were satisfied with their involvement in collaborative implementation, and what factors affected their ability to participate in collaborative implementation. Differences revolved primarily around collaborative history.

Collaborative partners on the Uncompahgre Project had a long history of collaboration with the local Forest Service unit and involvement in the planning of projects that were implemented using the CFLRP authority and funding. By contrast, partners on the Front Range Project were not involved in planning projects implemented under the CFLRP banner. Both groups of interviewees provided feedback on implemented treatments during collaborative group meetings and field trips, participated in monitoring, and generally provided feedback informally through conversations or scheduled meetings with Forest Service personnel. Additionally, participants of both projects felt that the Forest Service, to some degree, had changed prescriptions and adjusted implementation in response to feedback. However, participants on the Uncompahgre Plateau Project overwhelmingly reported satisfaction with their participation in implementation, while participants of the Front Range Project reported mixed levels of satisfaction and a desire for greater involvement beyond the opportunities they were able to utilize. Uncompahgre Plateau Project participants did not seek out additional opportunities for involvement beyond their initial, front-end collaboration, while Front Range Project participants attempted to formalize their involvement in implementation through their adaptive management strategy.

In both case studies, interviewees discussed several factors that impeded and facilitated the ability of stakeholders to participate in implementation in a way they found satisfactory. These factors have been organized into five main categories (Figure 11.1). Two categories are structural and individual factors, both of which are primarily associated with agency institutions (i.e. laws, policy, incentives, and agency-wide communication) and individual personnel who oversee implementation of policies in the field. Also relevant are factors that were entirely or partially external to the agency, including: collaborative characteristics, which we mean to encompass the role of agency

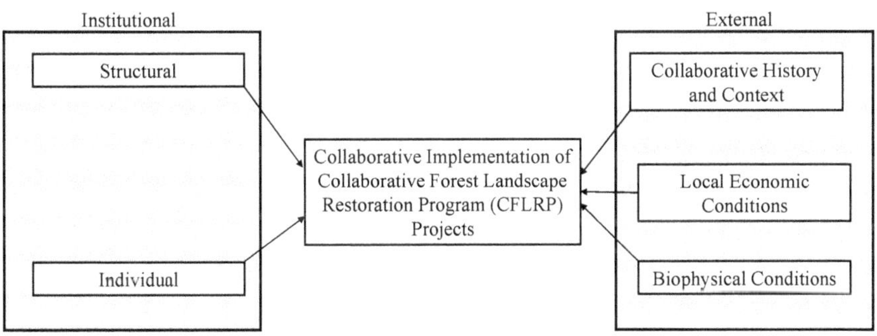

Figure 11.1 Factors Internal and External to the Agency Influencing Collaborative Implementation

partners in collaborative processes; local social, political, and economic conditions; and biophysical conditions. This framework for collaborative implementation, builds on the policy implementation literature and its emphasis on top-down and bottom-up approaches, while combining aspects of the conceptual models presented in Steelman (2010) and Moseley and Charnley (2014).

Structural factors, including the flexible requirements of CFLRP and constraints related to NEPA, affected engagement in collaborative implementation. In particular, collaborative participants viewed planning processes under NEPA as significantly influencing collaborative implementation. A project's decision under NEPA sets sideboards, or non-negotiable elements, for implementation, thus constraining the flexibility of treatments to accommodate adjustments proposed by collaborative participants. While this structural variable created some constraints, what really mattered was how groups navigated NEPA-related requirements. On the Front Range Project, stakeholders were only engaged post-planning, leaving them in a more limited and reactive position from which to shape treatment implementation. By contrast, on the Uncompahgre Project stakeholder involvement in planning, and the consequent enshrinement of agreed-upon project constraints and characteristics, left them more confident about agency implementation going forward. In other words, the structural factors were something to navigate, but could be a barrier or a facilitator to satisfactory collaboration, depending on the collaborative history and culture.

Individual factors were also relevant to explaining the differences we saw across these projects. On the Front Range Project, interviewees pointed to the reticence of individual agency personnel to collaborate, although it was unclear to them whether the problem rested with these individuals, their leadership, or an ambiguously defined 'agency culture'. However, they also noted that the culture on the two national forests they were working with was significantly different, indicating that problems and opportunities rested either with the local individuals they were working with, or local unit-level cultures. On the Uncompahgre Project, collaboration was a continuation of how business had been conducted for over twenty years. On this project, personnel on the national forest and collaborative partners had decades of experience working collaboratively and treated it as an assumed aspect of practice on that unit. The local line officer for the Uncompahgre Project, who at the time of our work had been on the forest for about ten years, also demonstrated a commitment to maintaining collaboration.

A key question is whether collaboration hinges on individuals, or whether a history of collaboration extends beyond the tenure of individuals to create collaborative capacity on a local unit and a local logic of practice that is shared by a unit and its partners. We suspect both are true. Within the broader culture, it is likely that the long history of collaborating on the Uncompahgre Project and forest created not just individual commitment from project champions, but also a local culture where collaboration throughout management processes became the norm and institutionalized in local practice, contributing to a greater satisfaction on a project that benefitted from collaborative implementation. Knowing this for certain would require additional research to determine the local logic of practice, a concept that is grounded in anthropological theory and often requires intensive ethnographic work to document. Our extensive experience with this project leads us to conclude, however, that the local culture and history with collaboration is something that is evident among many staff members, but which can also be facilitated or stymied by individual leaders and line officers, who in this case maintained and grew the local culture of collaboration.

We also observed that collaborative characteristics and history influenced trust and subsequent decisions. With the Uncompahgre Plateau Project, stakeholders discussed how they had built trust with the Forest Service over a long history of successful collaboration (probably both affinitive and rational trust; see Stern and Coleman in this volume). They expected the agency to implement according to the values and goals they had jointly identified. In this case, high levels of trust resulted in less involvement in implementation. However, the dissatisfaction with implementation, and the perception that some Forest Service personnel acted as barriers to collaboration, indicates that some participants of the Front Range Project trusted the Forest Service less. With less trust that the Forest Service will implement projects to the stakeholders' satisfaction, participants of the Front Range Project sought opportunities for greater involvement in the implementation process, such as developing a collaborative implementation strategy.

Local contractors, local capacity, and biophysical conditions also influenced project design, although our data did not reveal these to be stand-out themes for explaining the differences we observed. For instance, both sites faced capacity limitations and some complexity and disagreement about treatment parameters, but these were not the variables that interviewees highlighted during our interviews. Our findings indicated that the collaborative history between the agency and its local stakeholder groups was the most explanatory variable in defining the nature of collaborative implementation. Structural variables were significant, but could be navigated differently based on collaborative history.

For this reason, we build on previous frameworks for explaining policy implementation, but suggest the need to highlight collaborative history and culture as a separate and important category for analysis when looking at policies that involve mandates to collaborate with stakeholder groups. This category affects trust and local practice norms around collaboration. The role of individual units and implementers in policy implementation has a long history of study and is worth maintaining as a separate variable for investigation. However, in collaborative contexts, individual units and implementer activities are shaped by collaborative history and characteristics. In summary, collaborative group characteristics must be considered when examining the implementation and institutionalization of policy mandates that involve collaboration, and this broader framework (Figure 11.1) is useful for examining the suite of factors that influence implementation in such contexts.

Policy and practice implications

Our findings suggest several key policy and practice recommendations. First, public lands management agencies can maximize the potential for collaborative implementation of restoration projects by investing in new NEPA analyses and providing stakeholders with the opportunity to become engaged early in the planning processes. Trust through implementation is more likely on projects where stakeholders have been involved for longer periods of time, and have seen their recommendations reflected in planning documents that guide implementation. Moreover, the NEPA process and analysis can shape the ability of stakeholders to continue to have a role through implementation. When deciding on what types of projects to include in such a program, successful implementation may be more likely in places with a history of working collaboratively and where stakeholder participation in the NEPA planning process has been initiated early and continues throughout the life of projects.

Second, we suggest that units and individuals need to learn how to engage in effective collaboration and embed it within their cultures. With the degree of influence that line officers and other Forest Service employees have in translating collaborators' objectives in the implementation process, it is critical that collaborative projects implemented on National Forest land are staffed by employees who support collaboration and that this capacity be built over the long run.

The critical need for large-scale forest restoration means that projects are being conducted across the nation using a variety of policy tools and authorities. Collaborative groups and Forest Service officials, who developed strategies for implementing CFLRP projects in a collaborative manner, have established new precedents for what a collaborative restoration process can look like, and these lessons will persist over time. To this end, policy implementation frameworks, like the ones we used in this chapter, are helpful for evaluating and comparing cases to identify factors that influence the implementation of policies with mandates for collaboration and looking for opportunities to improve both policy and practice.

References

Ansell, C. and Gash, A. (2008) 'Collaborative governance in theory and practice', *Journal of Public Administration Research and Theory*, 18(4): 543–571

Aplet, G., Brown, P., Briggs, J., Mayben, S., Edwards, D., and Cheng, T. (2014) 'Collaborative implementation of forest landscape restoration in the Colorado Front Range', *Colorado Forest Restoration Institute*, Colorado State University, https://cfri.colostate.edu/wp-content/uploads/sites/22/2017/12/2014_Aplet-et-al.-FR-CFLRPAdaptiveManagement_CFRI-TB-1403.pdf

Berk, G. and Galvan, D. (2009) 'How people experience and change institutions: a field guide to creative syncretism', *Theory and Society*, 38: 543–580

Butler, W.H. (2013) 'Collaboration at arm's length: Navigating agency engagement in landscape-scale ecological restoration collaboratives', *Journal of Forestry*, 111(6): 395–403

Butler, W.H., Monroe, A. and McCaffrey, S. (2015) 'Collaborative implementation for ecological restoration on us public lands: Implications for legal context, accountability, and adaptive management', *Environmental Management*, 55(3): 564–577

Cheng, A.S. (2006) 'Build it and they will come? Mandating collaboration in public lands planning and management', *Natural Resource Journal,* 46(4): pp. 841–858

Cheng, A.S. and Randall-Parker, T. (2017) 'Examining the influence of positionality in evaluating collaborative progress in natural resource management: Reflections of an academic and a practitioner', *Society and Natural Resources*, 30(9): pp. 1168–1178, http://dx.doi.org/10.1080/08941920.2017.1295493

Cheng, A.S., Gutiérrez, G.J., Cashen, S., Becker, D.F., Gunn, J., Merrill, A., Ganz, D., Loquori, M., Saah, D.S., and Price, W. (2016) 'Is there a place for legislating place-based collaborative forestry proposals? Examining the Herger-Feinstein Quincy library group forest recovery act pilot project', *Journal of Forestry*, 114(4): 494–504

Cleaver, F. (2001) 'Institutional bricolage, conflict and cooperation in Usangu, Tanzania', *IDS Bulletin*, 32(4): 26–35

Colorado Forest Restoration Institute. (2008) 'Historic forest structure on the Uncompahgre Plateau: Informing restoration prescriptions for mountainside stewardship', Colorado State University, https://cfri.colostate.edu/wp-content/uploads/2016/05/2008_UncMesas_HistoricForestStructure1.pdf

Creswell, J. (2009) *Research Design: Qualitative, Quantitative, and Mixed Methods Approaches*. Thousand Oaks, CA: SAGE Publications

Cromley, C.M. (2005). 'Community-based forestry goes to Washington' In R.D. Brunner, T.A. Steelman, L. Coe-Juell, C.M. Cromley, C.M. Edwards, and D.W. Tucker (Eds.) *Adaptive Governance: Integrating Science, Policy, and Decision Making*. New York: Columbia University Press

Fernandez, S. and Rainey, H.G. (2006) 'Managing successful organizational change in the public sector', *Public Administration Review,* 66(2): 168–176

de Koning, J. and Bennecker, C. (2013) 'Bricolage practices in local forestry'. In B. Arts, J. Behagel, S. van Bommel, J. de Koning, and E. Turnhout (Eds.) *Forest and Nature Governance: A Practice-Based Approach*. New York: Springer

Lachapelle, P.R., McCool, S.F. and Patterson, M.E. (2003) 'Barriers to effective natural resource planning in a "messy" world', *Society and Natural Resources,* 16(6): 473–490

Lipsky, M. (2010) *Street-Level Bureaucracy: Dilemmas of the Individual in Public Service*. New York: Russell Sage Foundation

Maynard-Moody, S., Musheno, M., and Palumbo, D. (1990) 'Street-wise social policy: Resolving the dilemma of street-level influence and successful implementation', *Western Political Quarterly,* 43(4): 833–848

Moseley, C. and Charnley, S. (2014) 'Understanding micro-processes of institutionalization: stewardship contracting and national forest management', *Policy Sciences,* 47(1): 69–98

Nie, M. and Fiebig, M. (2010) 'Managing the national forests through place-based legislation', *Ecology Law Quarterly*, 37(1): 1–52

Sabatier, P.A. (1986) 'Top-down and bottom-up approaches to implementation research: a critical analysis and suggested synthesis', *Journal of Public Policy,* 6(1): 21–48

Schultz, C.A., Jedd, T., and Beam, R.D. (2012) 'The collaborative forest landscape restoration program: A history and overview of the first projects', *Journal of Forestry,* 110(7): 381–391

Singleton, R.A. and Straits, B.C. (1999) *Approaches to Social Research*. 3rd ed. New York: Oxford University Press

Sousa, D.J. and Klyza, C.M. (2007) 'New directions in environmental policy making: An emerging collaborative regime or reinventing interest group liberalism', *Natural Resources Journal*, 47: 337–444

Steelman, T.A. (2010) *Implementing Innovation: Fostering Enduring Change in Environmental and Natural Resource Governance*. Washington DC: Georgetown University Press

Williams, D.R., Stewart, W.P. and Kruger, L.E. (2013) *Place-Based Conservation: Perspectives from the Social Sciences*. New York: Springer

Wondolleck, J.M. and Yaffee, S.L. (2000) *Making Collaboration Work: Lessons from Innovation in Natural Resource Management*. San Francisco, CA: Island Press

12 Policy design to support collaborative landscape restoration

Lessons from the CFLRP

Courtney A. Schultz and Kathleen B. McIntyre

Introduction

A key question in the environmental governance scholarship is how to design policy to support governance approaches that promote collective action, balance stability of centralized institutions with flexibility for local participation, and match the spatial extent of management efforts with key ecological processes (Huitema *et al.*, 2009; USFS, 2015; Craig *et al.*, 2017; DeCaro *et al.*, 2017). The complex set of public demands on national forests, including biodiversity conservation, watershed protection, recreation, and commodity production, along with the many diverse landscapes and involved partner networks, make US forest management a rich environment for investigating this question. Innovative policies like the Forest Landscape Restoration Act (P.L. 11–111; 16 U.S.C. §7303), which created the Collaborative Forest Landscape Restoration Program (CFLRP), provide a unique opportunity to study in real time how policy changes might lead to governance transformations and allow actors to overcome longstanding barriers to change. In 2017, we conducted a programmatic review of the CFLRP to understand whether it led to the desired outcomes of improved collaboration and increased pace and scale of action on priority landscapes. In this chapter, we share our findings on the benefits of the CFLRP approach to collaborative implementation, examine some of the factors that led to variability across project sites, and discuss suggestions for future policy design.

The need for policy change to support collaborative landscape restoration

Since the 1970s, scholarship has emphasized the importance of better matching the scale of management actions to those of ecological processes of interest (Imperial, 1999; Cash and Moser, 2000; Cumming *et al.*, 2006). In response to this need, US forest managers in the 1990s emphasized that they were embracing ecosystem management approaches which focused on planning across longer spatial extents and longer time frames, with a greater emphasis on land management outcomes in addition to commodity outputs. Concurrently, community-based forestry groups demanded greater space for participation with the US Forest Service in defining the objectives of forest management in specific contexts (Cortner and Moote, 1999; Cromley, 2005; Maier and Abrams, 2017). Over time, collaborative approaches have become increasingly common on national forests, with numerous groups working on restoration of natural processes across large landscapes (Schultz *et al.*, 2012; Moseley and Winkel, 2013).

Despite these developments, for decades there had been little change to formal institutions in national forest management, and by the early 2000s the need for such change was becoming apparent. Prior to the passage of the CFLRP, diverse actors organized to tackle the problem of restoring natural fire regimes through efforts such as the US Fire Learning Network and the Western Governors' Association Forest Health Advisory Committee (Butler and Goldstein, 2010; Schultz *et al.*, 2012). From these efforts, a consistent message emerged: policy change was necessary to support collaborative restoration at an increased pace and scale.

Prior to the CFLRP, the institutions that defined the roles of collaborators and shaped project selection and design were out-of-step with the realities on the ground (see the Foreword and Introduction to this book for more background). A particularly pressing challenge was creating space for collaboration in the context of existing legal mandates (Sousa and Klyza, 2007). While no one wanted to return to an era of special interest group influence, community-based forestry groups (i.e. place-based, multi-stakeholder groups) demanded a better avenue for participation than traditional notice-and-comment procedures (Cromley, 2005; Sousa and Klyza, 2007). Informal collaborative approaches and place-based efforts that lacked agency buy-in and flexibility to adapt to changing conditions had led to some less-than-ideal outcomes, often with the Forest Service implementing only pieces of collaboratively designed projects, leading to frustration and broken trust (RVCC, 2009; Schultz *et al.*, 2012; Cheng *et al.*, 2017 and see the Introduction to this volume). Stakeholders were interested in moving towards greater involvement and improved accountability to collaboratively developed agreements; in essence, they were seeking more formalized co-management, or 'the sharing of power and responsibility between the government and local resource users' (Berkes, 2009, p. 1962). Federal partners also needed better avenues for leveraging the capacity of local partners, given that federal forest management relied more than ever on the resources of non-state partners to accomplish work and bridge across sectors and levels of government (Goldsmith and Eggers, 2004; Kettl, 2006; Abrams *et al.*, 2017).

Another major challenge was improving the coordination of activities across contiguous landscapes to affect fire behaviour. Restoration of most forested ecosystems in the western United States depends on fire as a key ecological process, and efforts must be coordinated across landscapes because of the spatial extent of natural fire events and the contiguity of fuels that drive fire behaviour (North *et al.*, 2009, 2012; Ryan *et al.*, 2013). Achieving this coordination requires the prioritization of investments in specific landscapes in order to affect fire behaviour. Prior to the CFLRP, the Forest Service had faced criticism from the US Congress for failing to effectively prioritize fuels reduction funding investments (US Senate Committee on Environment and Natural Resources, 2006). Policy documents written by the Forest Service to outline their strategy for making progress emphasized the need for 'focused investment' and partnerships in order to increase the pace and scale of restoration activities on national forests (USFS, 2015). In addition, to make restoration more affordable the agency recognized the need for better engagement of industry partners to reduce the cost of conducting work on the ground and to support utilization of low-value wood byproducts of restoration (USFS, 2010). The CFLRP was part of this strategy, and the outstanding question has been whether it has been a successful policy approach for supporting collaborative landscape restoration.

Policy design to support collaborative landscape restoration

The CFLRP was innovative as a policy tool in that it emphasized planning at a minimum spatial extent, required a landscape strategy for ecological restoration, and required that proposals for funding address social and economic considerations. It also included requirements for collaboration throughout all stages of the process. In addition, the CFLRP directed multi-year funding commitments to priority landscapes; funds could be used across a suite of restoration activities, instead of being tied to specific activities (e.g. fuel and vegetation management, habitat management, or monitoring) as is the case with traditional line-item appropriations. It was also unique as a nationwide program, with broad goals and national support, coupled with substantial flexibility to tailor project design to local contexts.

The CFLRP incorporates policy design principles that scholars recently have highlighted as valuable for improving governance, including, in particular, the notion that legal authorities should couple centralized support and enabling conditions with some flexibility to tailor approaches to local contexts (Craig *et al.*, 2017). This, ideally, allows partners the stability that might be provided by higher level institutions, along with the ability to tailor approaches to local conditions that may require specific activities, partnerships, or scales of action. Scholars have also noted that policies supporting participatory decision-making on landscapes with well-defined boundaries are more likely to facilitate collective action, which is needed to generate agreement, leverage resources, and work across jurisdictional boundaries (Lemos and Agrawal, 2006; Ostrom, 2009; DeCaro *et al.*, 2017). To be successful, policies also must give actors tangible support, such as funding or technical resources, and provide actors with the authority to make decisions (DeCaro *et al.*, 2017). In co-management contexts, collaborative monitoring, effective leadership, and bridging organizations or forums (i.e. opportunities to bring together multiple parties that have different skills and knowledge) also support collective action (Berkes, 2009). Through its collaborative mandate to work together from planning through implementation and monitoring, focused funding investments, and as a nationwide program with a suite of goals and requirements that could be tailored to local contexts, the CFLRP included many of these recommended aspects of policy design to support improved governance. The stage was set for improvements to practice facilitated by policy design.

Research design and methods

Our research focused on whether the CFLRP was leading to some of the desired changes in governance approaches (e.g. improved collaboration and scales of planning). Assessing the land management outcomes of the program was beyond the scope of our research. One indicator of success we considered was whether project participants were planning and implementing more restoration work at larger spatial extents. Because accomplishing monitoring has been a long-standing challenge in forest restoration, it was also important to look at whether CFLRP project participants were conducting monitoring and finding new ways to improve their efforts (DeLuca *et al.*, 2010; Biber, 2011). The program included social and economic goals as well. While these could be viewed as means to the end of restoration, they were also intended to build long-term social and economic capacity in resource dependent communities (Schultz *et al.*, 2012). The goals of the collaborative mandate were not specifically outlined in the law, but we assume the objective was to improve relationships to facilitate social agreement, learning and innovation, and the leveraging of resources; these too were important potential outcomes of the CFLRP to consider.

The outcomes of collaboration are multifold, making it useful to employ a typology for understanding the effects of a collaborative policy mandate. First-order outcomes of collaboration include trust and relationship building to improve collaborative capacity and develop collaboratively designed products to guide action (Innes and Booher, 1999; Ansell and Gash, 2008). Often these gains then yield beneficial 'second-order' effects, where the trust and informal institutions created through collaboration extend into arenas beyond the original venue of action (Innes and Booher, 1999). This might include the diffusion of collaborative capacity beyond CFLRP project areas or the development of innovative practices through a process of collective learning (Heikkila and Gerlak, 2013).

With these considerations in mind, the primary research objective for this work was to identify the extent to which the policy approach of the CFLRP led to desired outcomes. Specifically, we assessed whether more work was implemented at an increased pace and scale under CFLRP. To understand social and economic outcomes we examined whether the CFLRP model supported improved engagement of industry partners and whether it supported collaboration. On this last point, we considered multiple facets of collaboration, including improved relationships, the creation of products, diffusion of collaborative practices, and learning and innovation. Our second objective was to understand which factors interacted with policy to promote or impede success. Therefore, we investigated what factors supported success and the role of well-known barriers, including: bureaucratic processes, lack of agency capacity and leadership, or a lack of collaborative history (Wondolleck and Yaffee, 2000; Ansell and Gash, 2008). As a final research objective, we sought to gather ideas for improvements to the next iteration of policy design to support improved collaborative landscape restoration.

In 2017, we were contracted by the US Forest Service to undertake an evaluation of the CFLRP. To do this, we built upon lessons we had learned engaging with researchers, stakeholders, and the Forest Service over the years of CFLRP implementation to create a set of interview and survey questions. To get program-wide perspective, we sought to interview two agency-affiliated participants and two collaborative participants from each project, starting with project coordinators and triangulating suggestions from both agency and stakeholder-group coordinators for additional interviewees who could provide a broad perspective. We conducted eighty-nine interviews across all twenty-three CFLRP projects, with forty-two agency and forty-seven collaborative participant interviewees (numbers associated with individual interviewee quotes are included in the Findings section below). We recognize that a potential bias in our work is that individuals who were least satisfied with the CFLRP, and therefore not participating in projects, were outside of our sample. Questions focused on the value added by CFLRP, challenges, innovations, and variables that supported success. Our sample size was constrained by our capacity and funding but, with regard to the program as a whole, we reached a point of near-saturation on our questions, recognizing that additional interviewees may have provided some nuanced details about their experiences. All interviews were recorded, transcribed, and coded using Dedoose, an online, qualitative analysis platform, for expected and emergent themes based on literature and our interview guide. We wrote memos on key topics to analyse our excerpts (Glesne, 2011).

We also designed a survey to identify broader patters of successes and challenges across the projects and obtain generalizable data. Because of federal funding requirements and our short timeline, we could only survey agency personnel. Our target population included staff members and line officers with active involvement in a CFLRP project. The final population that we identified, with help from agency personnel, included 331 staff members and line officers (i.e. forest supervisors or district rangers, who have formal leadership responsibilities and decision-making authority on national forests). We had a 69 per cent response rate, with 229 survey respondents. We focused on descriptive information for each questionnaire item. In this chapter, we report frequencies, rather than means, for ease of interpretation and because our standard deviations were small. For more information regarding our research team, methods, and findings, please see our other publications on this project (Schultz *et al.*, 2017a, 2018).

Findings

Benefits realized under the CFLRP

In our survey, large majorities said they had moderate to strong levels of success of planning at landscape scales (83 per cent), integrating planning and activities across resources (85 per cent), and increasing the pace of restoration (75 per cent) (Table 12.1). Over two-thirds of respondents also reported moderate or strong success in accomplishing land management goals, such as restoring natural fire, reducing the threat of fire to local communities, and improving habitat, watershed function, and vegetation conditions. One NGO partner (#42) reported, 'My value is that it increases at pace and scale. It … help[ed] us get to a level where we're going to make a difference in the future health of the forest.' Over 60 per cent of survey respondents also said the CFLRP helped them improve monitoring and identify implementation efficiencies. One Forest Service staff member (#8) reported on the value of monitoring, stating, 'The ecological monitoring or the landscape monitoring that's funded through CFLRP has been invaluable in my opinion. I think for our particular collaborative, that's what brought a lot of trust in.' The exception in terms of land management activities was conducting prescribed fire, which almost all interviewees said they had not been able to accomplish at desired levels. In addition, interviewees on one project reported that not all the work that was getting done aligned with the collaborative group's vision.

In terms of economic outcomes, the CFLRP offered more mixed success. Interviewees said that where industry capacity already existed, the program supported contractors and led to the support or expansion of wood products facilities, with interviewees in several places saying the program had prevented mill closures, led to an increase in shifts, or supported new product development. One non-governmental organization (NGO) partner (#44) stated, 'It's been very helpful for our contracting partners and the mill

Table 12.1 Percentage of respondents reporting success in making progress on project goals (%)

Item	*Moderately Successful*	*Very Successful*	*Total*
Strengthen relationships with stakeholders	31	65	96
Plan at landscape scales	28	55	83
Integration across resource areas	35	50	85
Increase pace of restoration	32	43	75
Support wood products contractors or facilities	24	27	51

infrastructure … because there's a sustained program of work on the table that everybody can see and that is well-supported within the stakeholder community.' However, in those places without industry capacity, interviewees said that the CFLRP did not provide enough certainty or valuable product to re-engage industry. In our survey question about success in meeting accomplishments since receiving funding, supporting wood products facilities or markets ranked among the lowest categories of activities.

Regarding collaborative outcomes, our results indicated that the CFLRP was a valuable model for strengthening collaborative relationships, with 96 per cent of our survey respondents saying they had moderate to strong levels of success in this area (Table 12.1). A Forest Service interviewee (#12) reported,

> Getting all these stakeholders that normally would not ever talk to each other in one room on a monthly basis, talking and working together and seeing each other out in the field, hearing their side of the story. It's amazing – it's invaluable.

Many interviewees felt the program forced the agency to take collaboration seriously and engage participants in meaningful ways throughout a project. Almost every interviewee said collaboration had improved, although some stakeholders on a couple projects still were not completely satisfied with the quality of collaborative involvement. At least 70 per cent of survey respondents agreed that they had seen improved levels of trust, transparency, accountability, and communication (each of these was measured separately in our survey). In addition, 75 per cent said they had seen decreased conflict and 61 per cent said they had seen decreased litigation. One Forest Service interviewee (#5) said about their collaborators, 'It's really cut down our deliberation time, our objection process that we're in now, [and] litigation has been essentially non-existent, because of the collaborative.' In terms of actively participating in project planning, implementation, and monitoring, strong majorities of staff (i.e. over 75 per cent) said collaborators had influenced both planning and monitoring. Two-thirds of staff in our survey said collaborators had influenced implementation. Multiple groups developed detailed monitoring plans together, and several others crafted explicit zones of agreement about restoration principles.

People also reported some second-order benefits of collaboration, including spillover effects and innovations. Many interviewees said CFLRP legitimized collaboration and set an example for other projects in the forest. Interviewees on several projects said they will continue to collaborate on a larger scale focusing on forest-wide issues after CFLRP. On one project, referring to innovations in collaborative planning and monitoring, a Forest service staff person (#3) said:

> It's spilled over into other projects like the [large vegetation project] that the forest did. The local stakeholders were asking the forest to basically

> model the CFLRP type of process, as far as working on that large project. There's definitely been spillover there that's been helpful.

Survey respondents also reported finding innovations in multiple areas, including in how they engaged with community members and stakeholders (77 per cent) and landscape-scale planning (57 per cent) (Table 12.2). Two projects noted they were using more conceptual environmental impact analyses (i.e. National Environmental Policy Act or NEPA documents) focusing on zones of agreement to plan on larger scales and alter treatments, as they learned together through field trips and monitoring of their early implementation activities. One Forest Service line officer (#63) explained, 'It's much more of a conceptual NEPA. We don't look at every acre on the ground. We, in the general way, say, "Here's the areas we think we're going to be working in and treating", and it's been a shift in how we do business.'

Finally, many said that collaborative and multi-year funding commitments resulted in leveraged capacity and collective action, with some evidence of unique activities to bridge across groups of participants. One person said:

> You have a number of stakeholders … and all kind of aimed at a similar goal.… We are all working on CFLRP, but we're also taking all that information, those lessons learned, and applying to other projects, and getting other grants. And, just being able to scale up the work a lot more through the different partners that are working.

Agency interviewees noted that established collaborative partners were valuable for communicating about the project with the broader public through various strategies, including pub crawls and public broadcasting specials. People on a few projects mentioned they had started working with utility

Table 12.2 Innovations identified since receiving CFLRP funding (%)

Item	*Innovations occurred but it's too early to tell if they were successful*	*We identified successful innovations*	*Total*
Engaging with community members and stakeholders	7	77	84
Leveraging funding from partners or collaborators	12	58	70
Landscape-scale planning	11	57	68
Accelerating implementation of restoration activities	12	55	67
Designing monitoring	27	43	70
Identifying adaptive management approaches	26	38	64

providers to pay for work within their project boundary. Interviewees on one project said a subgroup worked to raise external dollars, bringing in over $10 million in grants. In our survey, nearly two-thirds of respondents said that since receiving CFLRP funding they had identified innovations in leveraging funding from partners or collaborators. Several stakeholder groups also were engaging unusual partners, including stakeholder-led local fire crews to complete burning, and local students to accomplish monitoring.

Facilitators and barriers to success

A key question for our work is what factors supported success, whether these were aspects of policy design or other variables. We also wanted to understand in more detail what factors interacted with the policy design of the CFLRP to allow some projects to be more successful than others. We considered projects to be successful when participants said they were on track to meet their goals and all interviewees were satisfied with both project progress and the quality of the collaboration.

Many people indicated that the policy design of the CFLRP was the fundamental variable supporting success. People attributed successes to the focused funding commitments of CFLRP, the emphasis on integrated restoration activities, and the flexibility to focus on multiple types of activities with the funding. Collaborators and agency staff also said that the CFLRP focused attention on high-priority landscapes, which brought in capacity and funding beyond the infusion of dollars from the program. An NGO partner (#41) noted, 'CFLRP provides some leveraging opportunity for additional funding and additional projects. When you have a number of stakeholders that are involved and all aimed at a similar goal … I think it's definitely been a powerful thing.' Explicit legal requirements to engage in monitoring, and funding flexibility to do so, also supported monitoring efforts – the most common area where stakeholders reported being involved in implementation and innovation. One Forest Service staff person (#1) explained, 'Historically, in the Forest Service, monitoring is talked about but never funded. So, I think that's a success, is actually getting the multi-party monitoring funding through the [CFLRP].' Stakeholders also said that the larger-landscape focus, multi-year funding commitment, and collaborative involvement gave their projects the social license to experiment with larger NEPA documents that encompassed more types of restoration work.

When we asked about the value of various aspects of the CFLRP approach in our survey, a few factors came out as being particularly important: 85 per cent of staff said the ten-year commitment of funding was very valuable, followed by the flexible dollars (69 per cent), and the requirements to work collaboratively (64 per cent) (Table 12.3). Support for the program as a whole was overwhelming; 91 per cent of staff surveyed and all interviewees said the program should continue. It is worth noting that these respondents were on funded projects, so we do not have agency-wide perspective on this issue.

Table 12.3 Survey respondent evaluation of different aspects of the CFLRP (%)

Item	*Somewhat Valuable*	*Very Valuable*	*Total*
The 10-year commitment of funding	11	85	96
The flexibility of the funding	25	69	94
The requirement to work collaboratively	28	64	92
The requirements to conduct monitoring	35	47	82
The competitive process for selecting projects	40	40	80

Also, about a third indicated that the prioritization of CFLRP projects and funding had negative effects on other programs or forests, simply because the program redirects funding to priority landscapes and away from other activities.

A number of our interview and survey questions asked what other variables supported or impeded success under the CFLRP. The most relevant variable was collaborative history. In our survey, aside from receiving funding, among the highest-ranked factors that supported success was the agency's history of collaboration with their stakeholders, with 91 per cent of respondents reporting this as somewhat or a major contribution to project success. Likewise, 88 per cent of respondents said that their stakeholders' history of working together was a factor underlying success, and among all external variables (i.e. factors external to the Forest Service organization), this was ranked as most valuable. Interviewees confirmed these findings. One Forest Service staff person, (#5), when talking about what had worked well for their group, said, 'The project was already in formation as a collaborative group, prior to CFLRP.... We didn't have to go through the process of developing a collaborative group and growing pains associated with that.' An NGO partner (#36) explained that they can call up their Forest Service partners to solve problems out of the public eye, commenting, 'The reason I can do that is I build relationships, and they trust me. They've seen that I'm not out to get them.' Some interviewees indicated that having a history of collaboration allowed them to tackle more contentious issues over the life of their projects.

Regarding facilitators and barriers related to the Forest Service, our findings emphasize the importance of strong leadership and adequate capacity. On multiple projects, people said that effective Forest Service line officers, or project coordinators engaged in strategic planning, took steps to make sure capacity was in place to support projects and created a culture of collaboration on their forests. One NGO partner (#46) reflected on the agency leadership and collaborative spirit stating, 'The forest leadership powers that be are very much behind those projects. They're not just paying lip service [to us] by any means. They really believe in the work and the need for it.' Some people also told us that local line officers redirected existing staff to projects, or lobbied the regional office of the Forest Service for needed capacity, but that this took strong leadership across all levels of the agency. People on successful projects said adequate capacity and, in particular, a full-time CFLRP coordinator was critical for planning, entering agreements, and implementation. Interviewees also said that having a facilitator, a science partner, and generally having scientific or legal capacity among stakeholders contributed to their ability to find agreement successfully. One Forest Service staff person (#18) reflected on the capacity provided by stakeholders stating, 'Between the science and folks on this forest, [we] were ready to move forward. Then we brought [an NGO] in early on to start analysing our landscape to help us prioritize things.'

Conversely, when leadership and available staff capacity was limited, this was a major challenge for projects. In our survey, the highest ranked internal barrier to success was staff turnover, with 77 per cent saying this was somewhat of a major barrier. Stakeholders said staff turnover undermined trust and slowed down projects. Some groups developed strategies to deal with this problem, including onboarding documents (i.e. documents to inform incoming staff members about project context) and meetings, but stakeholders consistently highlighted staff and line officer turnover as a major, ongoing challenge. One Forest Service staff person (#2) said,

> As an example, … six of the eight [interdisciplinary] team-level positions that we need for planning these landscape-level projects … have turned over two or more times. So, we just don't maintain the continuity that we need to keep the vision going.

In addition, we heard that in some places the agency did not match the focused investments under the CFLRP with adequate capacity to operate at an increased pace and scale. Aside from staff turnover, lack of capacity across multiple areas was identified by our survey respondents as the most significant factor within the agency that limited project success.

Two other external variables affected project success. Perhaps the most significant challenge involved the lack of capacity on the industry side. There were several instances of contractors going bankrupt during the tenure of their contracts, and some lacked the capacity or experience to work at the increased pace and scale needed. In our survey, the top external factor hindering project success was a lack of wood products, contractors, or markets, with 60 per cent of respondents marking this as a factor that hindered success. Meanwhile, several project areas faced disturbance events, including major wildland fire events or insect outbreaks. These events made planning documents and zones of agreement obsolete. A few interviewees noted that when these types of events occurred they forced partners and the agency to reconsider restoration goals, and sometimes redirected limited capacity towards responding to these disturbances.

Desired future changes

On a final note, we asked people what should change about the CFLRP program. The primary recommendation was to ensure some funding for maintenance of treatments. In the survey, 84 per cent of people said they would not be able to maintain their restoration treatments once the CFLRP funding ended, raising significant concerns about lost returns on CFLRP investments. A couple of people also suggested staged funding, with some seed money to support agreement building and planning prior to receiving a major investment of dollars. Many people supported the idea of allowing for a wider range of restoration activities under the CFLRP. Several collaborative participants also recommended incentives for staff to stay in place longer, opportunities to reward innovation, and improvements to accountability and communication with stakeholders (e.g. about annual accomplishments or work plans).

Discussion

The requirements to focus investments in particular locations, the multi-year funding commitments, and the requirements for collaboration were successful aspects of the CFLRP policy design. Our findings indicate that multi-year, federal funding commitments for priority landscapes supported collective action by signalling to collaborators that investing their time and resources in such places was worthwhile. Multi-year funding that is flexible, which can be used for a variety of activities, supported restoration across multiple resource areas on priority landscapes, something that we have found to be true of similar Forest Service changes to appropriations structures (Schultz *et al.*, 2017b). The funding model of the CFLRP also supported increased monitoring, which is important in co-management contexts and for restoration, but historically has been difficult to accomplish (Berkes, 2009; DeLuca *et al.*, 2010; Biber, 2011). Requirements to focus on a defined landscape for multiple years allowed for certain locations to be prioritized for restoration; this landscape focus also comports with evidence in the literature that well-defined spatial boundaries support collective action (Ostrom, 2009). For these reasons, we conclude that collaborative requirements coupled with multi-year, flexible funding commitments to specific landscapes is a policy approach that can effectively support collective action, prioritization, and landscape planning.

Regarding collaboration, we considered whether this policy model supported more effective co-management, with a greater formal role for stakeholder groups to participate in project planning, implementation, and monitoring – despite the legal and administrative constraints we outlined in the introduction to this chapter. In the early years of the program, it was unclear how the Forest Service would navigate the challenges of maintaining its legal responsibilities to the broader public and requirements to collaborate with an organized group of stakeholders under the CFLRP (Butler, 2013). As researchers, we wondered whether the language of the CFLRP would give stakeholders enough of a formal role in project design and implementation. Among our interviewees and in our survey results, the overwhelming trend was that people were satisfied with the CFLRP model. To our surprise, stakeholders did not tell us they wanted more formal authority under the law, although some stakeholders said they wanted more transparency from their Forest Service partners regarding budgets, work plans, agency policy, and project pre-planning. For these reasons, we conclude that the CFLRP was a successful model in that it allowed for some degree of co-management, supported improved collaboration, and sustained partner involvement in almost every case. However, while it is possible CFLRP works for now, we also note that national forest governance has increasingly moved towards greater formal levels of co-management. As this trend continues, it may be necessary to explore additional ways to provide formal rights for stakeholders in national forest governance, or to explore options around other legal constraints that may prove to be out-of-step with the realities of collaborative restoration and other activities moving forward.

In terms of policy change, two variables require attention going forward. One is the problem of industry capacity and the markets for restoration byproducts, which often are of low or no value. The CFLRP vision was that, by making a significant amount of product available, industry partners might materialize. However, stakeholders and agency interviewees alike told us that where industry did not already exist, the CFLRP did not provide enough certainty for significant new business capacity to relocate to the area. Some partners also said that the amount of product was offered for only ten years was not certain enough for them to expand their businesses significantly. We concluded that more attention needs to be paid to the industry side of the problem of restoration, and this may be an area for future policy development.

In addition, agency leadership and policy makers must identify ways to support the initiation and maintenance of landscape-focused projects. Opportunities may be needed to build collaborative capacity and begin planning in places that are priority locations for work, but not currently poised to participate in a program like the CFLRP, which offers funding only for implementation of treatments and requires the presence of collaborative partners (see 16 USC 7303 (f)(1)). In addition, people on every project we spoke to said they would need additional investment going forward to ensure the maintenance of treatments so that restoration investments can yield long-term benefits. This may require attention to funding options for continuing to invest in CFLRP project areas after CFLRP formally ends.

Policy and practice implications

To support future collaborative restoration efforts through policy design, the multi-year, flexible funding commitment to priority landscapes is a successful model for promoting collective action and a coordinated program of work across space and time. Partners brought capacity and connected the agency to science providers (see Colavito in this volume) and to members of the public. In addition, compared to the place-based policy approach, the general requirements of the law to work collaboratively with both agency and stakeholder buy-in on a broad set of goals at a minimum spatial extent were effective in supporting a multitude of projects that could be designed to meet local conditions, while also meeting national goals and having national-level support. This struck the important balance between local flexibility, coupled with certainty and stability at higher-levels of governance that the scholarship suggests is necessary for improving environmental governance (Folke *et al.*, 2007; Craig *et al.*, 2017). In addition, although this is beyond the scope of this chapter, we also noticed that having multiple projects operating under the same program allowed for projects to compare experiences through regional and national meetings among community-based forestry groups to support information sharing and learning.

For future project selection, when choosing priorities for investment, our findings, like those of other researchers, suggest that places with a history of collaboration are likely to be more successful at collaborating going forward (Ansell and Gash, 2008). However, where such a history does not already exist, particularly if places are priorities for restoration, the agency and policy makers could consider whether and how to invest in building collaborative capacity. In addition, while prioritization is valuable for accomplishing restoration, it also has its trade-offs. In other contexts, Forest Service personnel have told us that for some resources, like endangered species, important work may not be in a priority landscape, implying that some resource work may be compromised in a system of spatial prioritization (Schultz *et al.*, 2017b). A challenge for the agency and appropriators will be finding ways to incentivize investments in priority landscapes while also leaving some flexibility for other important resource work. Although this was not a major theme in our interviews, over the course of the CFLRP people have raised the question with us about the consequences of creating 'winners and losers'. In an era of declining resources for the Forest Service, and when the agency is focused on restoring large-scale ecological processes like fire (see e.g. 36 CFR 219 et seq., also known as 'the planning rule') prioritization is a necessity, but the consequence is that some places and resources will receive less investment. This begs the question, however, as to whether some resources and locations may be systematically undervalued or may get stuck in a losing cycle.

A major theme throughout our research was that agency practices needed to be better aligned with the emphasis on prioritization and collaboration under the CFLRP. To support collaborative efforts at the local level, turnover, particularly among those in leadership positions, needs to be minimized or handled more effectively. This might require opportunities to be promoted in place or incentives to stay in locations with active collaborative projects funded through focused investments. It is also essential to ensure that those in leadership positions throughout the agency are champions of collaboration and they create a supportive culture among their staff. This is especially true in co-management contexts where leadership is critical to success (Berkes, 2009). In addition, to support prioritization, the agency will need to figure out how to move capacity to priority locations, which will require a shift in practices from previous eras. Put simply, the agency must provide adequate staff capacity to support priority projects.

References

Abrams, J., Huber-Stearns, H., Bone, C., Grummon, C., and Moseley, C. (2017) 'Adaptation to a landscape-scale mountain pine beetle epidemic in the era of networked governance: the enduring importance of bureaucratic institutions', *Ecology and Society*, 22(4)

Ansell, C. and Gash, A. (2008) 'Collaborative governance in theory and practice', *Journal of Public Administration Research and Theory*, 18(4): 543–571

Berkes, F. (2009) 'Evolution of co-management: role of knowledge generation, bridging organizations and social learning', *Journal of environmental management*, 9(5): 1692–1702

Biber, E. (2011) 'The problem of environmental monitoring', *U. Colo. L. Rev.*, 83(1)

Butler, W.H. (2013) 'Collaboration at Arm's Length: Navigating Agency Engagement in Landscape-Scale Ecological Restoration Collaboratives', *Journal of Forestry*, 111(6): pp. 395–403

Butler, W.H. and Goldstein, B.E. (2010) 'The US Fire Learning Network: springing a rigidity trap through multiscalar collaborative networks', *Ecology and Society*, 15(3)

Cash, D.W., and S.C. Moser. (2000) Linking global and local scales: dynamic assessment and management processes. *Global Environmental Change*, 10: 109–120.

Cortner, H.J. and Moote, M.A. (1999) *The Politics of Ecosystem Management*. Washington DC: Island Press

Craig, R., Garmestani, A., Allen, C., Arnold, C., Birgé, H., DeCaro, D., Fremier, A., Gosnell, H., and Schlager, E. (2017) 'Balancing stability and flexibility in adaptive governance: an analysis of tools available in US environmental law', *Ecology and Society*, 22(2)

Cromley, C.M. (2005) *'Community-based forestry goes to Washington', in Adaptive governance: Integrating Science, Policy and Decision Making*. In R.D. Brunner, T.A. Steelman, L. Coe-Juell, C.M. Cromley, C.M. Edwards, and D.W. Tucker (Eds.). New York: Columbia Press

Cumming, G.S., Cumming, D.H.M., and Redman, C.L. (2006) 'Scale mismatches in social-ecological systems: causes, consequences, and solutions', *Ecology and Society*, 11(1)

DeCaro, D., Chaffin, B., Schlager, E., Garmestani, A., and Ruhl, J.B. (2017) 'Legal and institutional foundations of adaptive environmental governance', *Ecology and Society*, 22(1)

DeLuca, T.H., Aplet, G.H., Wilmer, B., and Burchfield, J. (2010) 'The unknown trajectory of forest restoration: a call for ecosystem monitoring', *Journal of Forestry*, 108: 288–295

Folke, C., Pritchard Jr., L., Berkes, F., Colding, J., and Svedin, U. (2007) 'The problem of fit between ecosystems and institutions: ten years later', *Ecology and Society*, 12(30)

Glesne, C. (2011) *Becoming qualitative researchers: An introduction*. White Plains, NY: Longman

Goldsmith, S. and Eggers, W.D. (2004) *Governing by network: The new shape of the public sector*. Washington DC: Brookings Institution Press

Heikkila, T. and Gerlak, A.K. (2013) 'Building a conceptual approach to collective learning: lessons for public policy scholars', *Policy Studies Journal*, 41: 484–512

Huitema, D., Mostert, E., Egas, W., Moellenkamp, S., Pahl-Wostl, C., and Yalcin, R. (2009) 'Adaptive water governance: assessing the institutional prescriptions of adaptive (co-) management from a governance perspective and defining a research agenda', *Ecology and society*, 14(1)

Innes, J.E. and Booher, D.E. (1999) 'Consensus building and complex adaptive systems: A framework for evaluating collaborative planning', *Journal of the American planning association*, 65(4): 412–423

Kettl, D.F. (2006) Managing boundaries in American administration: the collaboration imperative, Public Administration Review, 66 (s1)

Lemos, M.C. and Agrawal, A. (2006) 'Environmental governance', *Annual Review of Environmental Resources*, 31: 297–325

Maier, C. and Abrams, J.B. (2017) 'Navigating social forestry – A street-level perspective on National Forest management in the US Pacific Northwest', *Land Use Policy*, 70: 432–441

Moseley, C. and Winkel, G. (2013) 'Sustainable forest management on federal lands in the US Pacific Northwest: making sense of science, conflict, and collaboration', In P. Katila, G. Galloway, W. de Jong, P. Pacheco, and G. Mery (Eds.), Forests Under Pressure: Local Solutions to Global Issues. Online at: www.iufro.org/science/special/wfse/forests-pressure-local-responses/

North, M., Stine, P., O'Hara, K., Zielinski, W., and Stephens, S. (2009) 'An ecosystem management strategy for Sierran mixed-conifer forests' Gen. Tech. Rep. PSW-GTR-220 (Second printing, with addendum). Albany, CA: US Department of Agriculture, Forest Service, Pacific Southwest Research Station.

North, M., Collins, B.M., and Stephens, S. (2012) 'Using fire to increase the scale, benefits, and future maintenance of fuels treatments', *Journal of Forestry*, 110: 392–401

Ostrom, E. (2009) 'A general framework for analyzing sustainability of social-ecological systems', *Science*, 325(5939): 419–422

RVCC (Rural Voices for Conservation Coalition). (2009) '2010 Appropriations issue paper: managing our public forests and rangelands for healthy, resilient ecosystems and a strong economy'. On file with authors

Ryan, K.C., Knapp, E.E., and Varner, J.M. (2013) 'Prescribed fire in North American forests and woodlands: history, current practice, and challenges', *Frontiers in Ecology and the Environment*, 11(1)

Schultz, C.A., Jedd, T., and Beam, R.D. (2012) 'The Collaborative Forest Landscape Restoration Program: a history and overview of the first projects', *Journal of Forestry*, 110(7): 381–391

Schultz, C.A., McIntyre, K.B., Cyphers, L., Ellison, A., Kooistra, C., and Moseley, C. (2017a) 'Strategies for success under Forest Service restoration initiatives', Ecosystem Workforce Program, Institute for a Sustainable Environment, University of Oregon, Eugene, No. 81

Schultz, C.A., Moseley, C., Mattor, K.M., McIntyre, K., and Ellison, A. (2017b) 'Key Findings and Recommendations Based on the USDA Forest Service Integrated Resource Restoration Pilot Third-Party Review', *Journal of Forestry*, 116(1): 5–12

Schultz, C.A., McIntyre, K.B., Cyphers, L., Kooistra, C., Ellison, A., and Moseley, C. (2018) 'Policy Design to Support Forest Restoration: The Value of Focused Investment and Collaboration', *Forests,* 9

Sousa, D.J., and Klyza, C.M. (2007) 'New directions in environmental policy making: an emerging collaborative regime or reinventing interest group liberalism?', *Natural Resources Journal,* 47: 378–444

USFS. (2010) The 4 forest Restoration Initiative: Promoting ecological restoration, wildfire risk reduction, and sustainable wood products industries: A proposal for funding under the Collaborative Forest Landscape Restoration Program. pp. 24 Available online at www.fs.fed. us/restoration/CFLR/2010proposals.shtml

USFS. (2015) The Rising Cost of Wildfire Operations: Effects on the Forest Service's Non-Fire Work.

US Senate Committee on Energy and Natural Resources. (2006) 'Implementation of the Healthy Forests Restoration Act', 109th Cong., 2d sess., S.Hrg. 109–506

Wondolleck, J.M. and Yaffee, S.L. (2000) *Making collaboration work: Lessons from innovation in natural resource management.* Washington DC: Island Press

13 Commentary on collaborative implementation

Richard D. Margerum

For many years there has been a sustained focus on collaborative approaches to natural resource management, but researchers have paid less attention to collaboration during implementation. My research on collaboration in the United States and Australia led me to focus on this issue because I discovered that many participants were frustrated with a lack of implementation after spending years deliberating, building consensus, and reaching agreements. Many of the participants viewed consensus as the goal, when the real challenge was to translate consensus into an ongoing approach to deliver results (Margerum, 2011). Thus, I am pleased to see the focus on implementation in this book, and all three of these chapters provide important contributions to this topic by building on theory, using a range of sound research methods, and identifying clear policy contributions. In this commentary, I highlight several common themes and questions raised by these chapters in relation to collaborative implementation.

All three of these chapters highlight that collaboration during implementation requires a different set of strategies and tools. Plans and agreements often have limited value on their own, because the participants involved in implementation are faced with complex settings, changing conditions, and a myriad of options. Instead, there is a need for structures and systems under which the parties will share power and create the means to interact and make decisions collectively. For example, Butler *et al.* describe joint processes for prioritization and qualitative field reviews that allowed stakeholders to interact with Forest Service staff on how treatments were being implemented. Similarly, Bergemann *et al.* describe how a joint landscape restoration team created a roundtable that used field trips and meetings to participate in monitoring and adjust prescriptions and treatments.

These seemingly minor mechanisms to facilitate communication, feedback, and joint decision-making are important institutional adjustments to the network of decision makers (Agranoff, 2007, 2016). They help define the role of a collaborative group through formal and informal communication, interaction, and decision-making – something that I have found in my research to be critical in sustaining collaboration during implementation (Margerum, 2011). Trust and relationships among individuals are critical, but high turnover means that agreements about roles and responsibilities, plans for future decision-making, and even job descriptions are critical to sustain interaction.

In the context of the CFLRP, well-defined collaborative roles led to improved relationships and a number of broader benefits that were predicted by scholarship and have been confirmed through the research in these chapters. In their national survey, Schultz and McIntyre found that these efforts under the CFLRP have strengthened collaborative relationships. This was echoed by the case study findings in the chapters by Butler *et al.* and Bergemann *et al.* There is often little value placed on these kind of secondary outcomes (Innes and Booher, 1999) because they do not translate into 'results on the ground'. However, these cases demonstrate that improved collaborative relationships had direct benefits for improving decision-making and appear to have reduced administrative costs. The chapters noted that the collaborative efforts increased capacity, provided new information, and involved new people in monitoring. As Schultz and McIntyre note, in some cases they also had direct monetary benefits for the Forest Service because 'litigation has been essentially non-existent, because of the collaborative.' In an era of reduced federal agency resources, it is important to document and highlight these very real benefits to agency budgets. Collaboration often requires upfront costs that are easy to quantify, but the benefits are often longer term and harder to document.

Another issue raised in this section is the role of evaluation and monitoring. Monitoring strategies were utilized in the cases described by Butler *et al.* and Bergemann *et al.*, and the national survey by Schultz and McIntyre revealed that the CFLRP has led to innovative monitoring strategies. Importantly, these innovative efforts are focusing on a range of feedback loops and engaging a diversity of stakeholders in the monitoring responses. This provides more immediate feedback to implementation efforts, can help build trust in decision-making, and may help build support for the importance of monitoring itself. Having non-agency stakeholders asking, 'Are these treatments working?' helps make a case for investing in these efforts.

My research of collaborative groups has found that they often serve in unique roles of bringing together a wide range of perspectives and data. Often there was no other single entity that integrated this diverse set of information, and only the collaborative groups were approaching management holistically rather than through a narrower mission of a single agency (Margerum, 2011). When linked to the collaboratively designed roles and responsibilities to maintain accountability that I described above, these monitoring efforts created the ingredients for effective adaptive management. However, the collective nature of this approach also makes it vulnerable to the budget fluctuations and administrative whims of participating agencies. Creating mechanisms to commit to long-term monitoring and adaptive management funding could be one of the most important steps for the success of these efforts over time.

Another theme that emerges on the positive side of collaborative implementation is that it can be an incubator of innovation. The chapters in this section highlight some important innovations facilitated by the Collaborative Forest Landscape Restoration Program. The types of problems being addressed by this program are shared by many other issues facing public lands, and the lessons from these efforts may provide an important policy and administration template for other place-based collaborative implementation efforts. All three chapters in this section highlight how participants in these processes learned from each other and learned how to be better collaborators. Schultz and McIntyre note that CFLRP not only helped legitimize collaboration but also helped incubate new collaboration on other projects. A key contribution of research from these efforts is to share such learning and innovation in collaborative forest restoration to other projects, programs, problems, and policies.

Despite these positive outcomes associated with CFLRP and collaborative implementation, the chapters in Part III highlight that there continue to be constraints on the ability of these collaborative implementation efforts to work effectively. Butler *et al.* highlight the constraints of the Interagency Standards for Fire and Fire Aviation Operations, while Bergmann *et al.* highlights the constraints of National Environmental Policy Act, although they note that effective collaborative groups find ways to navigate legal requirements. Several authors throughout this book highlight issues related to agency turnover, which Schultz and McIntyre point out as an issue across all of the CFLRP cases.

These constraints are not new and much research over the years has noted these same issues (Margerum and Whitall, 2004; Thomas, 2003; Wondolleck and Yaffee, 2000). I believe we should assume these constraints as given conditions. For decades, national standards, NEPA analysis, and agency turnover have created constraints, so the more relevant question is how collaborative programs can be designed to be flexible and adaptable to allow these constraints to be overcome. All three chapters in this section highlight some options for addressing these issues, while recognizing these questions need ongoing attention. How can agencies better support collaborative, large-landscape restoration efforts in the face of these issues? How can the innovative strategies being used in some of these cases be expanded to other places?

There is also a broader question raised by these cases related to the capacity and participation of non-government stakeholders. Schultz and McIntyre found that collaborative history was one of the most important factors for success. This was echoed in the findings from Butler *et al.* and Bergemann *et al.* who suggest that a history of collaboration is an important supporting factor. One broader question is how to address regions with low capacity. Do these areas become lower priority or do they require new types of investment to increase capacity? For all areas, an important question is the sustainability of current stakeholder engagement. Could increasing dependence on stakeholder input lead to burnout, decreased capacity, and limited diversity of participants?

In addition to the findings and themes noted above, there are several other key questions for future research. The important question that remains is whether the approaches are effective in achieving the desired social-ecological outcomes. Carefully crafted, long-term outcome monitoring is important for assessing these efforts. Ideally, this could be structured to include long-term assessment across numerous cases operating under similar and different types of management approaches to allow more comparative analysis. These three chapters suggest the high potential that collaborative implementation in this program holds, but for a more definitive answer we'll need to wait for more robust data to come over the ensuing years.

Furthermore, there is a need for more research on diversity of participants in these efforts. By this, I do not just mean environment and timber interests, but how well do these processes incorporate underrepresented communities who live, work, and play in these areas? Collaboration has often been biased towards those with the power, time, and resources to participate (Margerum, 2011). Rongerude and Sandoval (2016) assert that creating a more inclusive process requires new strategies for collaboration, which may mean moving away from the table, as the location and metaphor for collaboration, to the streets and places where more diverse populations can be encouraged to participate.

A key contribution of this volume is to translate theoretical findings into policy and practice implications. In addition to the policy implications raised in these chapters, I would highlight a central implication of this program for agencies. CFLRP projects are addressing complex problems with many unique and changing contexts. CFLRP is set up to provide long-term commitment and flexibility, which are often cited as keys to making collaboration effective. Therefore, the program is well-positioned to be a model for other kinds of efforts to encourage collaboration on the most pressing problems, while recognizing that it requires flexibility and a sustained effort.

Within such a context, there is tremendous value in incubating cross-collaborative learning among those engaged in these kinds of collaborative efforts. Agencies hiring new staff should expect them to be trained in collaboration and communication, and universities who fail to provide this training are failing to meet the demands of modern management. However, collaboration also requires on the job training, mentoring, and people with the experience of navigating the litany of barriers (Margerum, 2016). CFLRP could provide on-the-job training for staff who will be working on these kinds of initiatives because there is a history of lessons and experienced staff who can help mentor and train. Ultimately, some of the most pressing problems facing federal lands will increasingly require more collaboration rather than less. The extent to which agencies can help mentor staff and partners in these efforts, the more likely they will be successful.

References

Agranoff, R. (2007). *Managing within networks : adding value to public organizations.* Washington DC: Georgetown University Press. Retrieved from www.loc.gov/catdir/toc/ecip071/2006031171.html

Agranoff, R. (2016). The other side of managing networks. In R.D. Margerum and C.J. Robinson (Eds.), *The challenges of collaboration in environmental governance: Barriers and responses* (pp. 81–107). Northampton, MA: Edward Elgar Publishing

Innes, J.E. and Booher, D.E. (1999). Consensus building and complex adaptive systems: a framework for evaluating collaborative planning. *Journal of the American Planning Association*, 65(4): 412–423

Margerum, R.D. (2011). *Beyond consensus: improving collaborative planning and management.* Boston, MA: MIT Press

Margerum, R.D. and Whitall, D.R. (2004). The challenges and implications of collaborative management on a river basin scale. *Journal of Environmental Planning and Management*, 47(3), 407–427

Margerum, R.D. (2016). Theoretical perspectives on the challenges of collaboration. In R.D. Margerum and C.J. Robinson (Eds.), *The challenges of collaboration in environmental governance: Barriers and responses* (pp. 27–53). Northampton, MA: Edward Elgar Publishing.

Rongerude, J. and Sandoval, G.F. (2016). From the table to the street: strategies for building a more inclusive collaborative process. In R.D. Margerum and C.J. Robinson (Eds.), *The challenges of collaboration in environmental governance: Barriers and responses* (pp. 317–337). Northampton, MA: Edward Elgar Publishing.

Thomas, C.W. (2003). *Bureaucratic landscapes : interagency cooperation and the preservation of biodiversity.* Cambridge, MA: MIT Press. Retrieved from www.loc.gov/catdir/toc/fy035/2002071779.html

Wondolleck, J.M. and Yaffee, S.L. (2000). *Making collaboration work: lessons from innovation in natural resource management.* Washington DC: Island Press.

Conclusion – the future of collaborative forest restoration

Scholarship, policy, and practice

William H. Butler and Courtney Schultz

Collaboration in forest management in the United States and globally has become more than just a trend; it has become ubiquitous. Not only are place-based collaborative groups and community-based forestry efforts widespread, but collaborative groups are seeking to increase their impact, work across larger landscapes, and engage in peer learning with communities of practice (Wondolleck and Yaffee, 2000; Weber, 2003; Goldstein and Butler, 2010; Margerum, 2011; Dukes *et al.* 2011; National Academy of Sciences, 2016; Abrams *et al.*, 2017). The anecdotal, case study, and survey-based evidence overwhelmingly indicates the value of collaborative governance for leveraging resources, promoting sustainable management, and supporting participatory approaches for managing social-ecological systems. While we know that collaborative governance has many benefits, we also know that it is a complex approach that can lack the predictability of hierarchical bureaucracy, that the benefits are unevenly distributed across collaborative groups and ecological systems, and that working at larger scales and over longer periods of time adds further complexity to the mix. Thus, much remains to be learned regarding how to most effectively engage in collaboration at large landscape-scales and what the benefits and trade-offs of doing so are.

This book seeks to address some of the questions that remain in collaborative landscape-scale forest management. Since the passage of the legislation that created the Collaborative Forest Landscape Restoration Program (CFLRP), researchers included in this volume have been conducting case studies and programme-wide investigations to understand a range of dynamics associated with this programme, with far-reaching implications. The chapters herein have shed light on legal requirements, how groups address collaborative governance dynamics, how they use science and work towards adaptive management, and how to work together beyond planning through implementation and monitoring. Studying these aspects of collaborative forest management in the context of CFLRP is important, because there has been limited study of collaborative efforts that are sustained throughout planning, implementation, and monitoring phases over long periods of time.

In this conclusion, we synthesize lessons learned from the chapters herein and augment these with observations from a symposium that we convened while preparing this volume. In November of 2017, authors included in this volume and high-level managers from The Wilderness Society, the Ecological Restoration Institute, the National Forest Foundation, the Colorado Forest Restoration Institute, the US Forest Service Forest Management staff, and the Rural Voices for Conservation Coalition participated in a workshop hosted by Florida State University's College of Social Sciences and Public Policy to discuss lessons learned and policy implications of our collective work and experience. In this chapter, we draw on the insights gleaned from the book chapters as well as discussions from the symposium to synthesize the research and primary lessons learned from this multi-year large-scale collaborative restoration effort. Our aim is to simultaneously draw out implications for scholarship, and for policy and practice of collaborative environmental management that will not only apply to US forest management, but will also apply to collaborative public land management globally.

Collaborative environmental management scholarship

The CFLRP provides the opportunity to deepen and broaden scholarship on collaborative environmental management in several areas. In this section, we synthesize the research across the chapters in this volume to draw out insights on five key topics. First, the CFLRP offered a chance to more deeply examine the question of how to go about engaging in collaboration mandated by policy. Second, there was an opportunity to explore the role of interpersonal relationships and how they could be established and nurtured over time. Third, collaborative processes require funding, resources, and staff time, but also the collaborative capabilities of the participants themselves. The CFLRP offered a chance to examine what capacities were needed and how they could be maintained. Fourth, the elusive practice of adaptive management has been a long-standing conundrum in collaborative environmental management, and CFLRP cases allowed scholars to examine the use of science and the practice of adaptive management. And, fifth, with the requirement to engage in collaboration through implementation and monitoring, CFLRP provided a chance for researchers to contribute more explicitly to the questions that arise in collaborative implementation. This section of our conclusion addresses each of these key theoretical contributions of the research in this volume.

Mandated collaboration

One of the defining characteristics of CFLRP is that it mandates collaboration throughout planning, implementation, and monitoring while funding restoration projects on publicly managed lands. A question for the research community was how the legal mandate to collaborate would manifest in practice. Collaboration is often seen as most effective when it is emergent and not overly formalized, making a collaborative governance system more flexible, adaptive, and creative than a conventional hierarchy or top down system of management (Healey, 2006; Innes and Booher, 2010; Emerson and Nabatchi, 2015; Bryson, Crosby, and Stone, 2015). Thus, there is an inherent tension between the requirement to collaborate in policy and what makes collaboration successful and efficacious. A key dynamic that stood out in CFLRP is the relative flexibility and ambiguity regarding collaboration in the legislation, which allowed groups to determine how best to navigate collaborative dynamics in their respective places. As such, the collaborative groups were able to organize themselves, often based on their history of past collaboration and their place-based practice experience (see Bergeman *et al.* in this volume; Monroe and Butler, 2016). Too much ambiguity in collaborative groups sometimes led to challenges as they moved through the collaborative process (see Stern and Coleman, in this volume; Coleman, 2016; Monroe, 2016). However, if the legislation had been too constraining and rigid in defining what constitutes collaboration, then the risks would have been so great that local and regional collaborative groups may have had a difficult time adapting to the requirements of the law. Thus, the balancing act between formality and informality, clarity and ambiguity, flexibility and organization, and predictability and adaptability remains important in designing policy and programmes to support collaboration, without overly constraining how it is done. It means that we may have to be satisfied with uneven performances across a range of groups. However, according to a Forest Service internal report, CFLRP cases have significantly outperformed their counterparts in other forests over the life of the programme. Moreover, based on work in this volume (e.g. Butler, Monroe and McCaffrey; Shultz and McIntyre), despite numerous challenges and limitations, CFLRP projects have made progress toward increasing the pace and scale of restoration. Thus, the broader outcomes may be worth the trouble of living through some failures in some cases in these types of programmes.

Relationships

More deeply understanding the importance of relationships and how they fluctuate and function, to either grease the wheels or stall progress of collaborative forest governance, is a major contribution of the work in this volume. Multiple chapters point to the importance of creating trusting relationships to get things done in collaboration (Bergemann *et al.*; Christensen and Butler; Schultz and McIntyre; Cheng *et al.*; Toman *et al.*; Stern and Coleman, in this volume). Stern and Coleman provide a useful typology for understanding trust and its various sources, while providing insight into how the ecology of trust works within a collaborative process. Christensen and Butler highlight the fact that accountability in collaborative groups depends upon relational accountability, which motivates good behaviour supported by social interactions with others in the group. These are difficult findings to translate efficiently into practice, as they do not necessarily point to a way to institutionalize or systematize effective collaboration. We can put all the right organizational and process frameworks into place, but, still, much depends on the individuals and relationships between people within the collaborative group.

Nonetheless, the work in this volume revealed that there are processes and strategies in organizing collaboration that can foster stronger and deeper relationships among the participants. Relationships depend upon repeated interactions to form, strengthen, and deepen. As a result, collaborative groups need to be organized in a way that ensures repeated and regular interaction among stakeholders over time. The chapters in this book point to the importance of getting together with a focus on common tasks and shared experiences, to go into the field and evaluate ecological conditions, to come up with a monitoring program and implement monitoring protocols, to develop visions for restoration or desired future conditions, to articulate and implement an adaptive management program, to explore prioritization and conduct post-treatment evaluations, to conduct and communicate new science, and more. These common experiences and shared tasks brought people together and facilitated the development of relationships, trust, and relational accountability bonds (Stern and Coleman; Christensen and Butler, this volume). This does not mean that all of these interactions uniformly resolve conflict, build stronger rational or affinitive trust, create the reciprocal expectations that ensure follow through, or any other positive relationship-building outcomes. However, without these types of interactions, relationships are likely to be relatively shallow and there will be limited opportunities to demonstrate reciprocity, follow through, and shared commitment to forest restoration.

Another key finding was that participant turnover can alter the dynamics of trust and accountability, which is largely dependent upon interpersonal relationships in collaboration. Particularly where relationships are strong, trust is high, and relational accountability is robust, turnover can undermine progress. This issue can be a barrier to collaborative accountability, a challenge to success, a source of breakdown in affinitive or rational trust, and a disturbance that tests the resilience of collaborative groups (Christensen and Butler; Stern and Coleman; Schultz and McIntyre; Cheng *et al.*; Ryan and Urgenson; in this volume; Stern and Baird, 2015). The issue of turnover has not been sufficiently addressed in existing literature. The research in this volume points to the importance of examining how to most effectively navigate the inevitability of turnover as people retire, change organizations, move to another region, get promoted, or in various other ways pull back from their participation in a project or collaborative group. This will help inform the practice of collaboration with agencies and organizations in environmental management, which often require long-term engagement to see projects through and show progress in achieving ecological and social outcomes.

Collaborative capacity

Another contribution of this volume sheds light on the importance of and approaches to building capacity to more effectively engage in collaborative forest restoration (Ryan and Urgenson; Bergemann *et al.*; Colavito; Butler *et al.*; Christensen and Butler; Schultz and McIntyre, in this volume). Collaboration takes time, energy, and resources. These issues were not mitigated by the legislation that created the CFLRP. Funding for implementation is not the same as funding for collaboration. The process itself is resource-intensive. Time and funding resources were a major factor that either impeded restoration or built momentum for restoration. Project groups were able to overcome funding constraints by seeking grant funding or having organizational partners contribute cost-share (both in-kind and cash) to the restoration project and collaborative process (Cheng *et al.*; Butler *et al.*; Bergemann *et al.*; Schultz and McIntyre). However, most faced capacity limitations trying to engage in the collaborative process while continuing to make progress on a demanding program of work to achieve restoration goals. Another source of enhanced capacity came from bridging organizations such as ecological restoration institutes in the western states or ecologically oriented organizations such as the Tall Timbers Research Station and Land Conservancy in the east, which were able to provide science and monitoring capacity to support CFLRP projects. University and agency-based researchers, likewise, were able to contribute to CFLRP science and technical needs in many of the landscapes (Colavito, in this volume).

Collaborative capacity goes beyond having technical expertise or staff time and funding; participants must have the ability and some measure of know-how to collaborate effectively. It is really about having the capabilities to effectively engage in the collaborative process (Bryson *et al.*, 2015). Some collaborative capability was infused into collaborative groups by hiring professional facilitators, while others relied on the capabilities and experiences of skillful and committed collaborators in their midst. Enhancing collaborative capabilities was also achieved through the participation of boundary organizations such as regional coalition groups and informal regional networks (Christensen and Butler; Colavito; Cheng *et al.*; Schultz and McIntyre, in this volume). Organizations such as Sustainable Northwest, the Rural Voices for Conservation Coalition, and the National Forest Foundation hosted, facilitated, and participated in regional and national workshops that brought CFLRP landscape partners together from across collaborative groups. These workshops, network meetings, and peer learning opportunities facilitated sharing ideas, working on common tasks such as reporting and monitoring projects, and creating new regional work groups that could address specific needs that spanned collaborative groups. Participants were able to build their own collaborative capabilities in these forums which served a learning network function (Goldstein and Butler, 2010). Being part of a national level program and operating under a similar policy context facilitated this type of dialogue across the cases and at a national level. That level of support and organization is a key opportunity for building collaborative capabilities, not just ecological management capacity.

Adaptive management and science

The CFLRP also yielded significant lessons about the role of science and science-providers in collaborative processes and the strategies and processes for monitoring and adaptive management. There has been a long-standing debate about how science might be utilized and accessed in collaborative processes and whether science providers must act as neutral providers of information, assisting, but not participating directly, in collaborative efforts (Karl *et al.*, 2007; Innes and Booher, 2010). However, in the CFLRP we saw a much more fluid relationship among collaborators, scientists who often were working as collaborative partners or with bridging organizations, and the many agency scientists who also brought expertise to the table (Colavito, in this volume). Development and use of science within collaborative efforts varied from place to place and over time, depending on who was available to gather and interpret information, the degree of controversy, and the rapidity of change. Gathering information from multiple sources, deliberating about it, and, most importantly, based on the work in this volume, getting out into the field to look at conditions in person to discuss observations and relate these back to scientific information, were all part of the messy but often productive processes that were part of CFLRP projects (Butler *et al.*; Christensen and Butler; Ryan and Urgenson; and Cheng *et al.*, in this volume). We also learned that, while we saw progress and innovation, monitoring and adaptive management remained challenging to undertake (Schultz and McIntyre and Cheng *et al.*, in this volume). Having the flexibility to fund monitoring was critical, as was having stakeholders who were committed to seeing it happen. Sometimes monitoring yielded important insights that allowed for breakthroughs, but, in most cases, projects have yet to synthesize monitoring data to build findings into future project designs. Perhaps most challenging was creating processes to feed information from monitoring back into project design and implementation. Indeed, improving this feedback process as part of engaging in adaptive management requires vision, commitment, and navigation of often mismatched temporal dynamics. It is a challenge revealed throughout this book and one that continues to require creativity and advancement of practice going forward (Cheng *et al.*, in this volume).

Collaborative implementation

Finally, CFLRP research has allowed us to make significant progress on understanding the dynamics and the value added of engaging in collaboration through implementation. Two key aspects of the CFLRP are that collaboration through implementation was mandated and program funding was all but promised over a ten to fifteen year time frame. As a result, agency personnel and collaborative groups could approach participation in the CFLRP with a long-term perspective and commitment. This did not mean that there were no tensions between collaborative groups and the agency. On the contrary, there were a great many. However, as the chapters in this volume point out, the key issue was not to eliminate conflict or concern, but, rather, to find ways to navigate through those tensions and achieve shared forest restoration goals. The chapters here affirm scholars who have made the point that the lines between planning, implementation, and evaluation are rather blurred (Nakamura and Smallwood, 1980; Pressman and Wildavsky, 1984; Margerum, 2011). Collaborative groups generally had an indirect influence on implementation of projects, but in rather substantive ways (Butler *et al.*; Schultz and McIntyre, in this volume). They found ways to navigate tensions by drawing on their history of working together and a logic of practice that had emerged over time (Bergemann *et al.*, in this volume). They also benefitted when they were able to maintain or enhance trust, accountability and capacity to sustain their interactions, and improve their planning and implementation actions, even under changing conditions (Stern and Coleman; Christensen and Butler; Ryan and Urgenson, in this volume). This long-term endeavour also demonstrated that collaboration through implementation of large-scale landscape restoration projects all but requires an adaptive management orientation with a focus on learning and change (Cheng *et al.*; Colavito, in this volume). It was fundamentally important that the landscape projects and collaborative groups were designed to be adaptive to ensure that, even during implementation, they were not tied to rigid plans and designs, as new science and monitoring, changing landscape conditions, as well as changing social, economic, and collaborative group dynamics all arose unexpectedly.

Policy and practice

Equally important to our contributions to the theory of collaborative environmental management, the work in this volume and insights from practitioners and researchers involved in our discussions at the aforementioned symposium reveal important new pathways for policy and practice of collaborative landscape forest restoration. In this section, we outline some of our key findings to provide guidance on forest policy, suggestions for agency programme management and administrative practice, as well as ways to improve collaborative practice in stakeholder groups. Again, while these policy and practice implications are inspired by and oriented to a US forest management context, they remain applicable to any democratic context where government agencies are responsible for managing public lands that require restoration interventions and that rely on collaborative engagement and input.

Policy implications

There are several facets of the Forest Landscape Restoration Act that researchers and practitioners have found to be important for landscape forest restoration. First, the research demonstrates that the long-term commitment of funding for implementation and monitoring distinguished the CFLRP from many other collaborative efforts and was essential for these projects. Funding is an important element of capacity in undertaking collaborative forest restoration at large landscape-scales, but it is the long-term commitment of funding that has been hard to come by and which was so crucial for supporting success under the CFLRP (Schultz and McIntyre, in this volume). The availability of funding was a key motivating factor to keep stakeholders engaged and agency programs of work moving forward. The long-term commitment supported collective action among stakeholders with their agency partners, created the impetus to work through conflicts and uncertainty during the process, led to innovation, and leveraged diverse capacities. At the same time, limitations on the funding brought up some challenges. Some landscape groups had trouble finding cost-share or matched resources sufficient to meet programme requirements. And, the inability to use funds for planning and for collaborative process management constrained projects in some cases. However, by and large, the consistent and substantial investment in this programme was seen as one of its strengths.

Second, mandating collaboration through planning, implementation and monitoring, even in an ambiguous and general sense through the legislation, helped legitimize and support enduring collaboration. Requirements for transparency and inclusivity helped promote engaging a wide variety of interests and increased communication of all facets of projects between agency personnel and collaborative stakeholders. This principle was unevenly applied, with some groups being strategically inclusive of a range of interests, while others were more exclusive in their level of participation (Butler and Esch, in this volume). Nonetheless, while the legislation was relatively vague on what constituted collaboration, CFLRP participants did not suggest a need for more formal legal requirements. Participants generally appreciated the flexibility to design collaborative organizational structures that were adapted to existing relationships and a history of working together in their respective regions. And, a recurring theme throughout our research and symposium was that the mandate to collaborate, especially when coupled with a history of collaboration and a commitment from agency staff to engage with stakeholders in good faith, led to project successes on multiple fronts.

Despite overall successes, there may be utility in defining transparency and inclusion more clearly, at least at the administrative level, if not in legislation. Agency staff members at national or regional levels had a difficult time holding landscape projects to account for uneven application of collaborative principles and process designs. Inclusiveness was one of the key factors that led to complaints. Striking the balance between open forums with the risks of unbalanced representation and exclusive venues, with their increased efficiency but reduced legitimacy, was an uneasy act for some of the collaborative groups. Those who were most successful had clear membership rules and a framework for decision-making that was inclusive of diverse interests. Tensions around transparency often arose in relation to communication between agency personnel and collaborative group stakeholders. There were several cases where the Forest Service planned restoration work without clear communication with the stakeholder group. Many project participants desired more active, timely, and transparent venues for sharing of agency budgets, work plans, performance measures/targets, guidance, and pre-project planning activities. Agency personnel should be allowed (and perhaps implored) to communicate all actions directly to their collaborative partners as well as adhere to standards of public participation and involvement broadly. Simultaneously, open communication about collaborative group processes and decisions can enhance the accountability and legitimacy of the group in relation to the agency and external stakeholders who are unwilling or unable to commit to participating in the group. Recommendations for enhanced and more open communications, as well as increased clarity on how to be more inclusive of a range of interests, were a recurring theme at our symposium, which suggests that multiple participants on various projects observed there was some room for improvement.

Third, many participants and observers have advocated for an all-lands approach to restoration so that adjacent lands not under federal management could be included in restoration plans and projects, but CFLRP funding can only be used on federal lands. Participants in our research and symposium suggested that more could be done to identify how to leverage work on CFLRP and other federally funded restoration projects to facilitate coordination and support for work on adjacent lands. There are multiple authorities that can be used on non-federal lands that could be paired with CFLRP-like funding sources. Also, policy programs like CFLRP could be written to allow funding to be used on adjacent non-federal lands (indeed, the Joint Chiefs Landscape Restoration Partnership supports cross-boundary work across public and private lands with federal funding). The original authors of CFLRP sought to develop an all lands approach in the legislation but had to abandon this idea when congressional members needed a clear line of authority and accountability for the appropriation of federal funds. This challenge is difficult to resolve, but necessary given the increased emphasis on working with and leveraging the capacity of non-federal and non-governmental partners in land management.

Finally, the role of capacity building and boundary organizations in supporting collaborative restoration work was a common theme across several chapters and the symposium. This function may be important enough that support for, or clarification of, a role for boundary organizations should be incorporated into legislation. In particular, cross-collaborative learning, coordination, and facilitation of collaborative groups at the project level are all functions that can be supported through boundary organizations. Cross-collaborative networks, regional gatherings, national workshops, science support through restoration institutes, and peer-learning sessions have all been identified as important venues for building relationships, enhancing capacity, and fostering learning to enable more efficient and effective restoration. Promoting these functions with dedicated funding could be helpful in future iterations of forest restoration policy.

Administrative practice

Beyond the policy implications, there are several implications for how agencies might engage more effectively in collaborative forest restoration. A central theme of many chapters in this volume, as well as the focus of much conversation among CFLRP participants and our symposium participants, is that turnover has been a widespread and consistent challenge. Capacity limitations in the agency have likewise slowed the pace of work. Across projects, researchers found that project participants, including collaborative stakeholders and agency staff, cited inadequate capacity and staff turnover as significant challenges to working at the pace and scale envisioned under the CFLRP (Schultz and McIntyre, in this volume). Turnover of key personnel in both agency and stakeholder groups undermined collaborative momentum, trust, and agreement (Stern and Coleman; Christensen and Butler; in this volume). Researchers and symposium participants concluded that a program at the scale of CFLRP requires an updated 'business model' for the agency – one that moves high-quality leaders to priority projects and rewards them for effective collaboration, supports focused investments with adequate capacity, and limits personnel turnover during the life of projects. Given that transitions are inevitable over the lifespan of long-term projects, more effective approaches to ensure smooth transition and a consistent commitment to collaboration should be sought. These approaches might include integrating collaborative group participants into agency personnel processes, incorporating expectations for collaborative engagement into agency job descriptions and personnel evaluations, and broadening the number of agency participants involved so that turnover becomes less disruptive. However, these ideas require some adjustments in how many forests operate today. Some of our colleagues at the symposium, who have been involved in forest management for many years, noted, often reading from cherished and underappreciated syntheses on collaborative efforts from the 1990s, that these notions of updating the agency business model were not new. With nationwide implementation of efforts like the CFLRP that come with significant investment of public funds, we hope that our sustained messaging on the need to reorient agency practices and organization to support collaborative efforts on priority landscapes will receive renewed attention.

Relatedly, research on CFLRP reaffirms that leadership, coordination, and facilitation are essential elements of effective collaborative environmental management. Researchers consistently found that leadership and coordination were essential to effective collaborative and agency interactions to support restoration. Having a designated coordinator to serve as a liaison and agenda setter from the agency improved communication and interaction between agency and collaborators on many of the projects. Indeed, where coordinators were present, collaborative participants felt that communication was stronger and trust could be built more easily (Stern and Coleman, in this volume). Agency line officers (forest supervisors and district rangers) who were committed to collaboration helped champion the process, guide agency personnel to more robust collaborative involvement and communication, and steered agency units to take collaborative recommendations seriously. Where leaders changed frequently or were not committed to collaboration, communication and trust broke down (Ryan and Urgenson; Stern and Coleman; Christensen and Butler, in this volume). Given the important role of line officers in legitimizing collaboration, ensuring that collaborative recommendations are taken into account, motivating staff to participate in collaborative processes, and having leaders who are collaborative champions rather than naysayers can make all the difference in the success of collaborative forest restoration. In addition, third-party facilitation was generally effective at helping stakeholders work through conflicts and keep strong communication processes between agency units and collaborative groups. Moreover, strong facilitation was associated with enhanced trust-building, accountability, and collaborative capacity (Stern and Coleman; Ryan and Urgenson, in this volume). Seeking ways to support leaders who are champions of collaboration, coordinators who can effectively liaison, and facilitators who can guide processes would improve the efficacy of project implementation.

Another important matter relates to the enduring challenge of fitting collaboration into a legal system meant to insulate agency decision-making. While the ambiguity around defining collaboration in the CFLRP was a potential strength, much of the research shows that more guidance might be needed on how agency personnel can participate effectively, especially in relation to the NEPA and FACA. Some agency personnel were unsure about legal requirements related to collaboration and hesitated to fully engage in the process. Some stakeholders were uncertain whether legal concerns were valid or being used as an excuse to avoid deeper collaboration. Some agency personnel withdrew from collaborative interactions, likely unnecessarily, citing legal constraints as a reason. This created tension and uncertainty among collaborative groups (Butler, 2013; Cheng *et al.*; Christensen and Butler, this volume). Transparent guidance on this matter would support trust-building with collaborators and enable more effective agency participation in collaborative restoration processes.

Agency guidance could also be improved in relation to monitoring and adaptive management. A long-standing practice in the Forest Service has been to undertake monitoring of implementation activities to ensure that actions outlined in a project or plan were implemented. However, such monitoring does not determine whether the projects were effective in leading to desired outcomes. Effectiveness monitoring is needed for more robust long-term adaptive management. On some CFLRP projects, staff were unclear about the distinction between scientific research and monitoring and erred on the side of doing less monitoring to ensure they did not cross a line (this issue related to allowable uses of appropriations for the National Forest System). This issue could be resolved with greater clarity of where the line between effectiveness monitoring and research resides so that monitoring opportunities to inform adaptive management are not lost. We suggest simply clarifying that if monitoring is done to determine the efficacy of management, even when it uses scientific principles such as controls and replication, this is effectiveness monitoring and an allowable use of funds under programs like the CFLRP. Simultaneously, closing the loop in adaptive management has been an ongoing challenge in ecological restoration efforts. The research in this volume demonstrates that when groups develop specific strategies for incorporating monitoring data into decision-making, deploy a range of monitoring techniques of varying levels of sophistication, and involve a diverse array of collaborative participants and agency personnel in the process, there is a greater chance that they can in fact inform future decisions. However, this is an area that would benefit from more research and dissemination of practical strategies in a variety of collaborative forest restoration contexts around the world.

Finally, programs like the CFLRP that are innovative, aim for restoration at larger scales, and require collaboration with external stakeholders, should be monitored and studied to ensure effective external accountability and review. Accountability can be a burden, but as these projects experiment with new approaches to restoration planning, implementation, and monitoring, external review can offer several benefits, including transparency, evaluation, and trouble-shooting, under the programme. Learning and knowledge transfer can also be enabled through such processes. Moreover, when things are not going well on individual projects or programme-wide, lack of progress can be identified early and either addressed through capacity-building interventions or funding can be redirected to other areas. Third-party evaluations, or an ongoing role for a diverse advisory committee like the FACA committee convened in the early years of the CFLRP, could guide or accomplish some of these efforts.

Collaborative practice

The CFLRP research also generated practical implications for collaborative governance more broadly, beyond agency administrative practice. The research has helped clarify how collaborative groups operated and contributed to agency decisions and actions while also maintaining flexibility to adapt collaborative design and processes to conditions in particular contexts. The studies in this volume all focused on a subset of cases from which to draw insights related to collaborative environmental management on long-term and large-scale restoration projects. There are several key lessons that can help guide practitioners engaging in collaborative practice in contexts such as these.

Regarding turnover, the CFLRP cases largely demonstrate that this can be a significant constraint to effective collaborative governance, but given the timeline of these cases it is inevitable. Instead of trying to stop turnover from happening, practitioners began to figure out how to navigate changes in staff in the agency and participants in the collaborative groups by building transition processes and procedures so that the negative effects from turnover were reduced. Many of the cases provide examples of strategies like handover memos when staff changed, overlapping appointments when possible, workshops to 'onboard' new people to collaborative groups and their processes, and narrative framing to tell the story of the collaborative and clarify its importance and role in restoration planning and implementation activities. Collaborative groups would be well-served if members advocated for the importance of the group in collaborative restoration and sought greater participation as part of processes to welcome new employees to the agency when they will contribute to collaborative activities. There is room for more research on this question of how to more effectively navigate turnover.

Our research and workshop discussions also revealed the essential importance of transparency. Transparency relates to trust, accountability, enhanced capacity, collective decisions, adaptive management, and collaborative implementation. Collaborative members can play an active role in ensuring that transparency remains high or is improved when insufficient. Collaborative groups benefitted from having an agency-based liaison or a collaborative member whose responsibility was to ensure open communication and engage in frequent check-ins with agency personnel. Much of the consternation that occurred during implementation happened when collaborative group members felt that they were unsure of how the agency was receiving or using recommendations, or when agency participation in collaborative meetings waned. As long as there were clear points of contact, regular involvement, and open communication, the issue of transparency was resolved. Collaborative groups can mitigate concerns by programming in regular reporting sessions and involving agency personnel in committees and appointing liaison duties to specific members.

Improving capacity for collaborative restoration is a multifaceted enterprise, and a common theme that arose in the research and at the symposium was that boundary and bridging organizations served a key role in helping to expand the capacity of many of the collaborative groups. These organizations, such as the ecological and forest-focused Southwest Ecological Restoration Institutes, advocacy groups such as the Rural Voices for Conservation Coalition or the CFLR Coalition, and learning-oriented interventions of the National Forest Foundation and Sustainable Northwest, to name a few, played essential roles on learning and monitoring project and programme progress. These organizations were effective at creating peer-learning spaces and fostering networking opportunities as they ran webinars, organized and hosted conferences, synthesized information and lessons learned across CFLRP projects, and served as a clearing house for resources. In these functions, boundary organizations were described as creating 'safe spaces' to promote learning by allowing participants in network meetings and peer learning sessions to admit failures, assess peer projects and activities, brainstorm ways to resolve problems, and share efficiencies and best practices. They also helped address capacity gaps, such as designing and implementing monitoring plans drawing on their science expertise or offering facilitation or coordination capacity for collaborative groups.

Summary of practice implications

We note that the lessons here for policy implications, agency personnel, and collaborators are similar, but the remedies differ depending on where the actors sit. The key challenges are to deal with capacity limitations and turnover, ensure transparency, and engage boundary organizations to support learning at multiple levels. We note that many agency personnel and participants alike recognized and addressed these challenges. Committed funding for a nationwide programme made possible cross-project, regional, and national learning, setting the stage for greater collective learning to improve both policy and practice over time.

Going forward, we emphasize a few points in summary. First, we reiterate the need for government agencies in public lands management, overseeing programs in collaborative restoration, to orient business models to support focused investment, ensure adequate capacity and effective leadership, follow investments, and ensure treatment monitoring and maintenance on priority landscapes. Collaborators will continue to require creative strategies to cope with limitations as agency practices evolve and everyone faces the realities of limited resources. Second, our work found that the CFLRP was a valuable approach to collaborative landscape restoration that almost all research subjects said should continue, but with ongoing evaluation and improvement. Key features were the commitment to long-term funding of implementation and monitoring, a competitive proposal development process, and a requirement for collaboration throughout planning, implementation, and monitoring. Improvements could incorporate an all-lands approach and support additional, necessary capacity building, collaborative management, and project planning in new locations as necessary. Finally, both research and practice have revealed opportunities to support more effective implementation in the future. We support ongoing synthesis and evaluation as part of new policy interventions and suggest it is critical to make the most of innovations and inform the future of both policy and practice.

Future research

The CFLRP provided an extraordinary opportunity to examine collaborative forest restoration under a unified policy and programme. While the research in this volume addresses some key outstanding questions in collaborative environmental management, it also leaves some questions unresolved and opens new questions for further examination. We outline some key areas for future inquiry following, generally, the three primary parts of this book.

Based on Part I of this volume, research on the CFLRP has emphasized the importance of relationships in navigating the collaborative process. Processes of trust, accountability, and capacity-building can help or hinder collaborative processes in myriad ways regardless of the type of process design or organizational form the collaborative group may take. This work also identifies questions that remain in collaborative environmental management to determine what factors contribute to sustaining effective collaborative group functions and contributions to restoration over time. Social-ecological systems are dynamic and changing systems. They are shaped by changing ecological conditions and disturbances as well as political and economic trends and shocks. Working in such contexts over large landscapes and significant time scales requires collaborative groups to navigate these complex and uncertain dynamics. Building on research in this volume, collaborative resilience is an emergent area of work that could expand our understanding of key factors that contribute to the ability of collaborative groups to adjust to dynamic change and uncertainty, absorb shocks, and retain their basic structure and function (Walker and Salt, 2006). Key questions include the following:

- How can collaborative groups effectively navigate turnover and change among participants from lead government agencies and collaborative partners?
- How can groups balance the need for flexibility and dynamism while also following requirements that may arise in law around the composition of such groups so that they are inclusive of multiple interests?
- To what extent and how can collaborative groups adjust to countervailing political or economic forces?
- How adaptive are the groups and planning processes in the face of ecological disturbances, such as wildland fire that impact upon restoration plans and projects?
- How nimble are planning documents, agreements, and collaborative groups to allow changing course when new conditions arise, or monitoring points to a needed shift in strategies?

Related to this idea of adaptation and resilience and based on Part II of this volume, another line of work should continue to pursue research on the use of science and adaptive management in environmental governance. CFLRP projects have begun to demonstrate a more nuanced and multifaceted view of adaptive management in practice. Rather than meeting the ideal model of adaptive management, CFLRP projects and collaborative groups have laid the groundwork and demonstrated that adaptive management can be undertaken in multiple ways, from using qualitative field evaluations and rapid assessments which provide quick feedback (and, at times, near immediate responsiveness), to a more systematic design of adaptive management procedures and decision tools. This work portends several research questions that require further investigation, including:

- What governance approaches will facilitate the adaptive management feedbacks that are necessary to ensure management decisions and actions are achieving desirable outcomes in the face of uncertainty?
- What myriad tools and techniques contribute to the adaptive management cycle at multiple temporal and spatial scales?
- How can monitoring, particularly of project effectiveness, be designed and supported to inform future project design?

In Part III of this volume, we focused on collaborative implementation and overall programme success. It is clear that there are blurred boundaries between planning, implementation, and monitoring activities, how they relate to each other, and how they are undertaken in long-term and large-scale landscape restoration. While collaboration through implementation and monitoring in this context revealed how government agencies and collaborative groups can work together in restoration, it also revealed that partners often have an indirect influence. While they were able to influence plans and management actions in very real ways, collaborators were rarely directly involved in implementation – and they were kept at arm's length on many of the internal workings of agency-led implementation processes. The ideals of co-management, where landowners, land managers, and users of resources collectively manage social-ecological systems and undertake shared activities to build agreement, implement actions, enforce compliance, and monitor outcomes, remains elusive when involving public lands. Moreover, the research in this volume was confined to collaborative groups and agency project teams that were participating in the CFLRP. What we did not study (as of yet) was the reverberating effects this program had on other forests, areas of the agency, and in other organizations involved in forest governance. This brings up several questions about the scale and approach to forest governance and management going forward, such as the following:

- To what extent can agencies and organizations work together to achieve co-management across boundaries under existing policy contexts?
- What new policy designs might be required to scale-up and scale out forest restoration beyond existing management and administrative boundaries?
- What administrative and stakeholder capacities must accompany these new approaches to governance?

- What does it take to institutionalize a new way of doing things in forest management and what does that process look like? Is ten years enough for these projects to integrate new modes of collaborating, planning, monitoring, and implementing restoration projects that can outlast the CFLRP program and funding?
- How extensive is the reach and influence of this landscape scale approach? To what extent does this approach to planning, implementation, and monitoring extend to other projects that are not part of the program?

Finally, we leave several much broader questions for other researchers to answer. Ultimately the CFLRP was meant to change ecological conditions on the ground. The work in this volume focused on the governance aspects of the CFLRP, but much work remains to be done as to whether the restoration efforts yielded the expected benefits and whether there were any unexpected detrimental effects on ecological integrity. In addition, we noted from the outset that collaborative governance is an approach to democratic participation. To what extent do the effects of collaborating through public lands management diffuse into other aspects of society and create new or improved capacities? Finally, directly measuring the other social and economic benefits of these types of projects is something that was largely beyond the scope of our collective work. This is an area for future research to investigate the multiple outcomes of social, economic, and ecological sustainability, that were, ultimately, the intent of the CFLRP.

Concluding thoughts

As the CFLRP comes to the end of its first ten years, there is much to reflect on regarding the future of collaborative forest management. Lessons learned from this program have the potential to inform future forest management policy and practice as well as collaborative governance theory and practice. This collection seeks to capture some of these lessons learned and offer guidance for new policy directions, make suggestions for enhancing forest management practice, and offer refinements for collaborative governance.

Forest management continues to be refined worldwide. In the last 100 years, we have learned to more effectively model ecological systems and functions, wildfire behaviours, disturbance regimes, resilience characteristics, and more. We have slowly shifted from an economic outputs orientation to an ecosystems services and ecological orientation in management decision-making. We have also moved toward community-based forestry and co-management approaches globally, which has given rise to new forms of governance to engage in cross-boundary management. And yet, the vestiges of prior institutions, perspectives, and behaviours remain. Therein lies a central tension facing the future of forest management. It has become clear that the scientific management era ushered in by the creation of public lands management agencies and the firm belief in the separation of scientific decision-making from the political dynamics of policy making has come to an end. Perhaps it was always a myth of how things were supposed to work but never a true tale of how they actually worked. It is a persistent narrative, either way. Cross-boundary and collaborative forms of management have been gaining legitimacy and demonstrating their effectiveness (by and large) for the past forty years. However, the legal, regulatory, administrative, and behavioural change necessary to transform the forest governance system is still quite significant.

CFLRP marks a pilot experiment for a new way of conducting business in US forest management. Even so, it is nested within former modes of doing business. It sets up a framework for competitive funding for projects, mandates collaboration throughout implementation and monitoring, and articulates a vision for restoration that addresses ecological health, reduces wildfire risk, and revives rural resource dependent communities. And yet, collaborative groups and agency personnel ran into barriers of existing laws, procedures, and practices that inhibited their ability to engage effectively and productively with each other toward common cause. Despite these barriers, however, most groups were able to make significant progress in working together to achieve restoration goals. These groups were able to draw on various forms of trust, navigate multiple accountabilities, strengthen relational accountability, and build the necessary collaborative capacity to move forward. They were able to articulate and converge on restoration visions and desired future conditions, set up adaptive management strategies that could actually work, and draw on the expertise and guidance of boundary organizations to address complex challenges related to science, values, and politics. And, they drew on their histories of successful collaborative practice models, engaged in multiple facets of collaborative implementation, and navigated pathways to success to varying degrees across the country. This combination of attributes is not a recipe for success, but certainly holds many of the requisite ingredients and may, in fact, represent the beginning of new era in environmental policy that funnels capacity towards places where state and non-state actors are collaborating to find creative solutions to enduring challenges in environmental governance.

References

Abrams, J., Davis, E.J., Moseley, C., and Nowell, B. (2017) 'Building Practical Authority for Community Forestry in and Through Networks: The Role of Community Based Organizations in the West,' *Environmental Policy and Governance,* 27(4): 285–297

Bryson, J., Crosby, B., and Stone, M. (2015) 'Designing and Implementing Cross-Sector Collaborations: Needed and Challenging', *Public Administration Review,* 75(5): 647–663

Butler, W.H. (2013) 'Collaboration at arm's length: Navigating agency engagement in landscape-scale ecological restoration collaboratives', *Journal of Forestry*, 111(6): 395–403

Coleman, K. (2016) 'The Role of Trust in Natural Resources Management', PhD Thesis. Virginia Tech, Blacksburg, VA

Dukes, E.F., Firehock, K., and Birkhoff, J. (Eds.) (2011) *Community-Based Collaboration: Bridging Socio-Ecological Research and Practice.* Charlottesville, VA: University of Virginia Press

Emerson, K. and Nabatchi T. (2015) *Collaborative Governance Regimes.* Washington DC: Georgetown University Press

Goldstein, B.E., and W.H. Butler. (2010). 'Expanding the Scope and Impact of Collaborative Planning: Combining Multi-stakeholder Collaboration and Communities of Practice in a Learning Network.' *Journal of the American Planning Association,* 76(2): 238–249

Healey, P. (2006). *Collaborative Planning: Shaping Places in Fragmented Societies.* 2nd ed. London, England: Macmillan Press

Innes, J.E. and Booher, D.E. (2010). *Planning with Complexity: An introduction to collaborative rationality for public policy.* New York, New York: Routledge

Karl, H., Susskind, L., and Wallace, K. (2007). A Dialogue, Not a Diatribe: Effective Integration of Science and Policy through Joint Fact Finding. *Environment,* 49(1): 20–34

Margerum, R. (2011). *Beyond Consensus: Improving Collaborative Planning and Management.* Cambridge, MA: MIT Press

Monroe, A. (2016) 'Structuring Collaborative Implementation on US National Forests: How Formality and Inclusivity Influence Effectiveness in the Collaborative Forest Landscape Restoration Program' PhD Thesis. Florida State University, Tallahassee, FL

Monroe, A. and Butler, W.H. (2016) 'Responding to a policy mandate to collaborate: Structuring collaboration in the Collaborative Forest Landscape Restoration Program', *Journal of Environmental Planning and Management,* 59(6): 1054–1072

Nakamura R.T. and Smallwood, F. (1980) *The politics of policy implementation.* New York: St. Martin's Press

National Academies of Sciences, Engineering, and Medicine. (2016) *A Review of the Landscape Conservation Cooperatives.* Washington DC: The National Academies Press

Pressman, J.L. and Wildavsky, A. (1984) *Implementation: How great expectations in Washington are dashed in Oakland.* Berkeley, CA: University of California Press

Stern, M.J. and Baird, T.D. (2015) 'Trust ecology and the resilience of natural resource management institutions', *Ecology and Society,* 20(2): 14 www.ecologyandsociety.org/vol.20/iss2/art14/

Walker, B. and Salt, D. (2006). Resilience Thinking: Sustaining Ecosystems and People in a Changing World. Washington DC: Island Press

Weber, E.P. (2003). *Bringing Society Back In: Grassroots Ecosystem Management, Accountability, and Sustainable Communities.* Cambridge, MA: MIT Press

Wondolleck, J. and Yaffee, S.L. (2000). *Making Collaboration Work: Lessons from Innovation in Natural Resource Management.* Washington DC: Island Press

Index

Page numbers in **bold** denote tables, those in *italics* denote figures.

www.ingramcontent.com/pod-product-compliance
Ingram Content Group UK Ltd.
Pitfield, Milton Keynes, MK11 3LW, UK
UKHW010959180926
475701UK00004B/458
* 9 7 8 0 3 6 7 6 6 2 7 4 5 *